Felix Leopold Oswald

Physical Education

Or The Health Laws of Nature

Felix Leopold Oswald

Physical Education
Or The Health Laws of Nature

ISBN/EAN: 9783337026202

Printed in Europe, USA, Canada, Australia, Japan

Cover: Foto ©berggeist007 / pixelio.de

More available books at **www.hansebooks.com**

PHYSICAL EDUCATION;

OR,

THE HEALTH-LAWS OF NATURE.

FELIX L. OSWALD, M.D.

"The Laws of Nature proclaim themselves and are their own avengers."—Thomas Campanella.

NEW YORK:
D. APPLETON AND COMPANY,
1, 3, AND 5 BOND STREET.
1882.

INTRODUCTION.

"Audendum est, ut illustrata veritas pateat, multique a perjurio liberentur."—LACTANTIUS.

"If the right theory should ever be discovered, we shall know it by this token: that it will solve all riddles."—EMERSON.

SINCE the dawn of the Germanic Reformation, the history of science has been the history of a triumphant progress. The spell of supernaturalism is broken. The nations of the Caucasian race have awakened to a recognition of the fact that this world of ours is not governed by capricious ghosts, but by consistent laws, and that the study of those laws is a path to all the knowledge and happiness our nature is capable of attaining. A spirit of free inquiry is abroad; we have found that the sun of nature will ripen more truth in a single year than the light of mysticism in a century; branches of science which had withered to the core have again put forth leaves and flowers, and begin to yield abundant fruit.

With one strange exception. The science of health is still a barren tree. Our wealthy minds dwell in poorly tabernacles; the right method of promoting man's physical welfare seems to be one of the utterly lost arts. The machinery of our printing-presses, locomotives, and power-looms has been brought to a wonderful degree of perfection, while the mechanism of our own bodies is getting more and more out of joint. We have ascertained the weight of distant worlds and the component elements of rarely seen comets, while the theories on the right quantity and quality of our daily meals are still sadly uncertain. Our horticulturists raise vegetables that would have astonished the cabbage-gardener of Felix Sylla, while ninety per cent. of our children are so puny and scrofulous that the Spartan Gerontes would have drowned them in Eurotas to put them out of their misery. The presidents of our geographical, philological, and astronomical societies are high-priests of an almost schismless church, while in medicine there are nearly as many different systems as colleges. Chemistry has raised agriculture to the rank of an exact science; bee-keepers have learned to apply new methods, and to avoid old mistakes; cattle-plagues are getting less frequent, but the diseases of the human race still multiply from year to year.

And, still stranger, neglect—i. e., indifference—can by no means be alleged in explanation of this contrast. The culture of the *humanities*—if I may use that word in its literal sense—has been recognized in its full importance. The study of anatomy, biology, and medical chemistry has been pursued with equal zeal and success. In *those* branches of his art, the average country physician is far ahead of Galen and Hippocrates. The cupboard of the modern *paterfamilias* contains drugs which Avicenna never dreamed of in his physiology. The constant supervision of our children could hardly go further. We guard their health with all the care of a Grecian *systarchus*—only not with the same success.

INTRODUCTION.

And yet I believe that this anomaly can be fully accounted for. About a century after Romulus had laid the foundations of the most successful empire, an East-Indian fanatic made the most successful attempt to undermine the foundations of human happiness. His original purpose was laudable enough: he tried by meditation and prayer to "discover the origin of evil" —i. e., of the misery which mankind have brought upon themselves by violating the laws of Nature. But, failing to diminish those evils by tracing them to their true cause, he conceived an idea which has increased them a thousand-fold: the unhappy idea that life itself is a curse, the gift of a malevolent demiurgos,* a disease whose only remedy is death. Upon this principle he founded a religion, which he preached with the eloquence and more than the common success of an enthusiast, for at his death two million of his countrymen had become converts to the DOCTRINE OF ANTI-NATURALISM, and actually believed that this earth is a vale of tears;† that science, industry, and all worldly pursuits are utterly vain; and that a man's natural instincts are

---

* "Many a house of life
  Hath held me—seeking ever him who wrought
  These prisons of the senses, sorrow-fraught;
  Sore was my ceaseless strife.
    But now, thou builder of this tabernacle—thou!
  I know thee! Never shalt thou build again
  These walls of pain, nor raise the roof-tree of deceits,
  Nor lay fresh rafters on the clay;
  Broken thy house is, and the ridge-pole split.
  Delusion fashioned it!"
      (Edwin Arnold's translation of the famous summary of
        the "Tripitáka.")

† "The first truth is of *sorrow*. Be not mocked!
  Life which ye prize is long-drawn agony!"
                —Ibid., chap. vii.

his natural enemies.* A moral virus seems to possess the propagative power of a contagious disease; Buddhism spread like the pest, and in the first century of our chronological era the most terrible of all Oriental plagues crossed the Hellespont and fell like a poison-blight upon the Eden of the Mediterranean nations.

Whether the saints of a later Buddhism † have procured us an inheritance in the clouds, whether

<blockquote>
* "The second truth is <i>sorrow's cause</i>. What grief
Springs of itself, and springs not of desire?
Senses and things perceived mingle and light
Passion's quick spark of fire!"
  (Edwin Arnold's translation of the famous summary of the "Tripitáka," chap. vii.)
</blockquote>

† "The essence of the Catholic religion is the center dogma of Buddhism: the doctrine of the worthlessness of terrestrial life. With this difference only, that Christianity dates that worthlessness from the transgression of our apple-eating forefathers. This modification implied the fiction of a *liberi arbitrii indifferentiæ;* but it was required by the necessity of grafting the doctrine of Buddh upon the mythological dogmas of Judaism."—Schopenhauer, "Die Welt als Wille," vol. ii, p. 694.

The patristic dogmas are a mixture of anti-naturalism and supernaturalism—a worship of Buddh in the guise of Jehovah. The gods of Greece were the deified powers of Nature, the deified patrons of husbandry and science. Our gods are the deified enemies of Nature. Not the corrupted form, but the very essence of Hebrew-Buddhism doctrine, is anti-natural. Whole nations have tried to put its doctrines into practice. The result is the desert—physical, moral, and mental. The Protestants still ascribe those results to the apocrypha of Catholicism. The Catholics ascribe them to the heresies of the dissenters. The truth is, that those apocrypha and heresies have saved us from utter ruin. Heterodoxy is an attempt to naturalize Christianity. The experiments of the patristic era have proved that unmodified anti-naturalism must lead to madness and bankruptcy.

"It can not be an accident. Buddhism and Christianity agree in all the particular peculiarities that distinguish them from the numberless they have revealed a preternatural light, may be mooted questions; but it is certain that their dogmas have cost us three million square miles of our best earthly inheritance, and that their rule has obscured the light of common-sense for fifteen hundred years.

other religions of the world."—Böttger, "Vergleichende Mythologie," vol. i, p. 217.

So numerous are the resemblances between the customs of this system and those of the Romish Church, that the first Catholic missionaries who encountered the priests of Buddha were confounded, and thought that Satan had been mocking their sacred rites. Mr. Davis ("Transactions of the Royal Asiatic Society," vol. ii, p. 491) speaks of the "celibacy of the Buddhist clergy, and the monastic life of the societies of both sexes, to which might be added their strings of beads, their manner of chanting prayers, their incense, and their candles." Mr. Medhurst ("China," London, 1857) mentions an image of a virgin, called the "Queen of Heaven," holding an infant and a cross in her arms. Confession of sins is regularly practiced. Father Huc, in his "Recollections of a Journey in Tartary, Thibet, and China" (Hazlitt's translation), says: "The cross, the miter, the dalmatica, the cope, which the Grand Lama wears on his journeys, or if he is performing some ceremony out of the temple—the service with double choirs, the psalmody, the exorcisms, the censer suspended from five chains, and which you can open or close at pleasure—the benediction given by the lamas by extending the right hand over the heads of the faithful—the chaplet, ecclesiastical celibacy, religious retirement, the worship of the saints, the fasts, the processions, the litanies, the holy water—all these are analogies between the Buddhists and ourselves." And in Thibet there is also a Dalai Lama, who is a sort of Buddhist pope. After the theory, "*que le diable y était pour beaucoup*," was abandoned, the next explanation of the Jesuits was that the Buddhists had copied these customs from Nestorian missionaries. But a serious objection to this theory is, that Buddhism is at least five hundred years older than Christianity, and that many of these striking resemblances belong to its earliest period.—(Clarke's "Ten Great Religions," p. 139.) Compare "Works" of Sir W. Jones, vol. i, pp. 87, 168; ibid., vol. iii, p. 105; Mill's "History of India," vol. i, p. 49; Coleman's "Mythology of the Hindoos," p. 193; "Asiatic Researches," vol. vi, p. 271, and vol. vii, p. 40.

Their hell-fire and witchcraft dogmas deluged the earth with blood; the victims of the Crusades, witch-hunts, and inquisitorial butcheries * would cover a continent with corpses. Their anti-Nature dogmas † paralyzed industry, the study of natural science was superseded by the worship of miracles, rational agriculture was neglected, ‡ and the garden-regions of the Mediter-

---

* "They felt with St. Augustine that the end of religion is to become like the object of worship, and they represented the Deity as confining his affection to a small section of his creatures, and inflicting on all others the most horrible and eternal sufferings. . . . Persecution invariably accompanied the realization of these doctrines, and their normal effect upon the character was to produce an absolute indifference to the sufferings of those who were external to the Church. . . . In every prison the crucifix and the rack stood side by side."—Lecky's "History of Rationalism," vol. i, pp. 326, 354.

"Hæretici non solum excommunicari sed juste occidi possunt."—Thomas Aquinas, "Summa," vol. ii, art. iii.

"If a man believe in this saving of souls by faith, he must soon think about the means, and if by cutting off one generation he can save many future ones from hell-fire, it is his duty to do it."—(Rogers's "Recollections," p. 49.) But there is a deeper reason. A harmless religion can dispense with bloodshed. The Greeks and Romans found no difficulty in propagating their joyous Nature-worship. "Procul profani!" was the cry of the Eleusinian priests; they had more followers than they wanted. But the authority of an anti-natural creed can be supported only by violence. That support withdrawn, Nature will reassert her rights, more swiftly if aided by Science, more slowly but not less surely by the influence of an instinctive reaction against an abnormal condition.

† Luke xiv, 26; Matthew x, 34–37; Luke xviii, 22–30; Matthew x, 9, 10; Luke ix, 3, 24; Luke xiii, 51; Luke x, 4.

‡ "The Spanish Christians considered agriculture beneath their dignity. In their judgment war and religion were the only two avocations worthy of being followed. Some of the richest parts of Valencia and Granada were so neglected that means were wanting to feed even the scanty population remaining there. Whole districts were deserted, and down to the present day have never been repeopled. All over Spain the same destitution prevailed. That once rich and prosperous coun-

ranean peninsulas became hopeless sand-wastes.* Their anti-reason dogmas † crushed free inquiry; a total eclipse of common-sense and science followed like an unnatural night upon the bright sunrise of Grecian civilization. The government of the world was usurped by the

try was covered with a rabble of monks and clergy, whose insatiate rapacity absorbed the little wealth yet to be found. The fields were left uncultivated; vast multitudes died from want and exposure; entire villages were deserted."—Buckle's " History of Civilization," vol. ii, pp. 52, 57.

"Except in the parts occupied by the Moors, the Spaniards were almost totally unacquainted with irrigation."—Clarke's "Internal State of Spain," p. 116.

* "The fairest and fruitfullest provinces of the Roman Empire, precisely that portion of terrestrial surface, in short, which about the commencement of the Christian era was endowed with the greatest superiority of soil, climate, and position, which had been carried to the highest pitch of physical improvement, is now completely exhausted of its fertility; . . . *a territory larger than all Europe*, the abundance of which sustained in by-gone centuries a population scarcely inferior to that of *the whole Christian world at the present day*, has been entirely withdrawn from human use, or, at best, is thinly inhabited. . . . There are parts of Asia Minor, of Northern Africa, of Greece, and of Alpine Europe, where the operation of causes set in action by man has brought the face of the earth to *a desolation almost as complete as that of the moon;* and though, within that brief space of time which we call ' the historical period,' they are known to have been covered with luxuriant woods, verdant pastures, and fertile meadows, they are now too far deteriorated to be reclaimable by man, nor can they become again fitted for his use except through great geological changes, or other agencies over which we have no present or prospective control; . . . another era of equal improvidence would reduce this earth to such a condition of impoverished productiveness as to threaten the depravation, barbarism, and perhaps even the extinction of the human species!"—G. P. Marsh, "Man and Nature," pp. 4, 43.

† John xx, 29; Matthew v, 3; John v, 24; 1 Cor. xiv, 37; John vi, 47; Galatians i, 8, 9; John viii, 51; 1 Cor. iii, 11; John xiv, 6; 1 Cor. xvi, 22.

priests of the new religion, ministers of Gehenna, who distributed the poison of pessimism, and fattened upon the spoils of the dead. Mankind slept in a fever-dream, and a swarm of vampires sucked their life-blood with impunity. Knowledge became a contraband; human freedom and human reason lay prostrate and fettered at the foot of the cross.

But there was one science which could not be neglected with safety. Here and there the sleepers began to awaken from their torpor, and the reign of anti-naturalism could be maintained only by constant coercion—by a constant and merciless suppression of the protests of outraged Nature. Rebellion had to be crushed in the bud, and the training of the young was made a preparatory school of slavery and idiocy. The resources of human ingenuity and of inhuman cruelty * were exhausted to devise an effective educational system for the *denaturalization* of the human race, for the suppression of the instinct of freedom,†

---

* "If any sect," says Ludwig Börne, "should ever take it into their heads to worship the devil in his distinctive qualities, and devote themselves to the promotion of human misery in all its forms, the catechism of such a religion could be found ready-made in the code of several monastic colleges."—Compare Llorente, "History of the Inquisition," pp. 129–142; "Codex Theodosianus," lib. ix, cap. 1, 2; ibid., lib. xvi, cap. 10; Palmer, "On the Church," vol. i, p. 13; Motley's "Rise of the Dutch Republic," vol. ii, p. 155; Llorente, pp. 273–275; Rohrbacher, "Histoire de l'Église Catholique," tome xvii, p. 210; Bédarride, "Histoire des Juifs," pp. 16–20; Buckle's "History of Civilization," vol. i, p. 500; Wachsmuth, "Der Bauernkrieg," vol. i, chaps. i–iii.

† "All the slavish submission ever exacted by the caprice of pagan tyrants was now brought into a system."—Circourt, "Histoire de l'Espagne," p. 282.

"The more a man was taught the less he would know; for he was taught that inquiry is sinful, that intellect must be suppressed, and that credulity and submission were the first of human attributes."—Buckle's "History of Civilization," vol. ii, p. 74.

of the love of truth,* of Nature, † of health,‡ of beauty,# of mirth, ‖ of earthly happiness,^ of the

---

* "The fathers laid it down as a distinct proposition that pious frauds are justifiable and even laudable, and, if they had not laid this down, they would nevertheless have practiced it as a necessary consequence of the doctrine of exclusive salvation. Paganism was to be combated, and therefore prophecies of Christ by Orpheus and the Sibyls were forged, lying wonders were multiplied and ceaseless calumnies poured upon those who, like Julian, opposed the Church. This tendency triumphed wherever the supreme importance of these dogmas was held. Generation after generation it became more universal, it continued *till the very sense of truth and the very love of truth were blotted out from the minds of men*."—Lecky's "History of Rationalism," vol. i, p. 395.

† "It was, moreover, wrong to take pleasure in beautiful scenery; for a pious man had no concern with such matters. . . . On Sunday it was sinful to walk in the fields or in the meadows, or enjoying fine weather by sitting at the door of your own house."—Buckle's "History," vol. ii, pp. 305, 313.

‡ "Bathing, being pleasant as well as wholesome, was a particularly grievous offense; and no man could be allowed to swim on Sunday. It was, in fact, doubtful whether swimming was lawful for a Christian at any time, even on week-days, and it was certain that God had on one occasion shown his disapproval by taking away the life of a boy while he was indulging in that carnal practice."—Ibid., vol. ii, p. 312.

"As bathing was a heathenish custom, all public baths were to be destroyed" (by order of the Spanish clergy), "and even all baths in private houses."—Ibid., vol. ii, p. 44.

"It was improper to care for beauty of any kind."—Ibid., vol. ii, p. 306.

‖ "Even on week-days, those who were imbued with religious principles hardly ever smiled, but sighed, groaned, and wept. . . . One pious elder had acquired distinction by his faculty for what was termed 'a holy groan.' He used to weep much in prayer and preaching; he was every way most savory. Even among young children, from eight years old upward, toys and games were bad; and it was a good sign when they were discarded."—Ibid., vol. ii, pp. 304, 305.

^ "A Christian must beware of enjoying his dinner, for none but the ungodly relished their food. By a parity of reasoning it was wrong for a man to wish to advance himself in life, or in any way to better his condition."—Ibid., p. 313.

confidence in the trustworthiness of our natural intuitions.*

Hence the difficulties of the social reformer. While other sciences had been merely neglected, *the science of human nature had been sedulously perverted.* Elsewhere we found only vacant halls, here a strongly garrisoned bastile. In arts, industries, and the objective sciences, the path of former cultivators could still be traced through the neglected fields; the progressive educator had to force his way through formidable obstacles.

The field of education has ceased to be an ecclesiastical allodium, but before we can cultivate the soil we have to uproot a jungle of poison-weeds. Our medical, physiological, and dietetic theories are still interwoven with countless prejudices. The antagonism of mind and matter is still an established dogma. The rapid progress of scientific discovery has not furthered the study of human nature, because man is still treated as an *alter ens*, a being governed by laws apart from, or even opposed to, those of Nature in general. An appeal to supernatural agencies, whose inefficacy in veterinary surgery has long been conceded by all but idiots, is still a favorite expedient in the treatment of human diseases. Anti-naturalism is still more prevalent. Nature and our natural instincts are still supposed to con-

---

* "According to this code, all the natural affections, all social pleasures, all amusements, and all the joyous instincts of the human heart were sinful. . . . The clergy looked on all comforts as sinful in themselves, merely because they were comforts. The great object of life was to be in a state of constant affliction. Whatever pleased the senses was to be suspected. . . . It mattered not what a man liked; the mere fact of his liking it made it sinful. Whatever was natural was wrong."—Buckle's "History of Civilization," vol. ii, pp. 312, 314. (Supported by a vast array of original quotations.)

spire for the destruction of our health and happiness. Sweetmeats, fruits, cold water, and fresh air, whatever recommends itself to our innate senses, is regarded with suspicion; repulsiveness and healthfulness are still synonyms. A list of "staple medicines" is a list of staple poisons. With a large class of medical practitioners alcohol still ranks as a remedial agent, and even as an article of food. It is well known that children and animals detest the smell of tobacco and the taste of brandy, coffee, tea, and pungent spices, but the significance of that aversion still remains unheeded. Our day of leisure is still the dreariest day in the week; the welfare of the soul is still supposed to be incompatible with earthly pleasures. We have a thousand mythology-schools for one gymnasium; the importance of physical culture, the interdependence of soul and body, and the moral influence of health, have hardly begun to be realized. Even in the dog-days parents still think it necessary to torture their children with woolen under-garments and greasy-made dishes. Scrofula, consumption, and the evidence of our noses have not yet taught us that fresh air is preferable to prison-smells. Millions of homes are still afflicted with the curse of the night-air superstition. In short, we have not yet outgrown the *prejudice against the trustworthiness of our natural instincts,* the baneful result of the natural depravity dogma. There is no hope of recovery while our daily food is such a poison, no chance of progress while our feet are tied with such shackles. The specters of the middle ages will not vanish till the Science of Life has been freed from the fallacies of anti-naturalism.

The object of the present volume is to indicate the most mischievous of those fallacies, and to suggest the best means of renaturalizing our system of physical education. I lay no claim to the merit of having extended the path of knowledge; I have merely dared to pursue it to its legitimate end. I have neither augmented nor improved the weapons of free inquiry, but I have tried to demonstrate the possibility of applying them to what I consider their most important purpose. For health is more than a means of happiness; it is happiness itself. Disease has no consolation but the hope of recovery. Salvation means *healing*. For a dyspeptic Eurystheus the apples of the Hesperides come too late. Without health, no earthly acquisitions are worth their price; our superlative spectacles are a poor substitute for the superlative eyes of the ancients. "Mental perfection" might be a compensatory acquirement, if it were not for the circumstance that only a healthy body can be the temple of a healthy mind. A mere toothache will bias the thoughts, the sentiments, and even the principles of the wisest man, as surely as a small fissure will mar the music of the best flute. And a weakly man is half sick. Physical vigor is the basis of self-reliance. The health of an effeminate person is a flame that can not weather a storm. And the chief problem of life would be solved if we could regain the longevity of our forefathers. Life would again be worth living. The seed of our youth would have time to yield a harvest; a laborer might hope to enjoy the fruit of his toil, and the vague yearnings after a *post-mortem* existence in the clouds * would vanish with the doubts about the value of the present

---

* "Agnosticism" (*skepsis*) "is by no means incompatible with the belief in the *possibility* of a future existence, nor with the poetry of that belief. Nothing can be franker or manlier than the meditations of Socrates on the threshold of an unknown world, or the still grander allegory of the Edda-mythus: 'No mortal can know what Odin whispered in the ear of his son, when Baldur mounted the funeral-pile.' "—Gotthold Lessing.

world. Health, strength, and long life, were once the free gifts of our All-mother, and we may yet regain our forfeited birthright, but we can not hope to be reconciled to Nature till we cease to worship her enemies, till we cease to consult the oracle of a life-hating fanatic, and to disregard the teachings of our life-preserving instincts. When we begin to heed the consistent revelations of the God of Nature, rather than the contradictions of his alleged deputies, we shall find on this side what we sought beyond the grave—we shall regain our earthly paradise.

"*Audendum est sapere.*" Let us dare to use our eyes. The night is giving way. The obscurantists have protected their dogmas by treating every torchbearer as an incendiary, but at the approach of dawn they shall ring their fire-bells in vain.

Felix L. Oswald.

Cincinnati, *October, 1881.*

CONTENTS.

CHAPTER I.

DIET.

Dietetic abuses the principal cause of human degeneration—Physiological effects—Redeeming influences—Hygienic regeneration—The safeguards of Nature—Simple tastes of children—Their study as a dietetic criterion—Stimulants—Their repulsiveness to normal men—Infantine diet—Milk—Best substitutes—Weaning—Wet-nurses—The nurses of the Pays de Vaud—Argyll peasants—Ionian Islanders—Baby-food—Xenophon's recipe—Hygienic precautions—Man-food—Haller's definition—Natural, artificial, and anti-natural diet—Common errors—Peptic stimulants—Spicing and cooking—Concentrated food—Objections to —Biological conjecture—Perfect health—Uncooked food—Experience of French soldiers—Vegetable diet—Its great variety—Semi-animal food—Eggs, milk, and butter—Cheese—Objections to—Brewer's yeast—Fastidiousness—Inconsistencies of dietetic purists—Best food for various climates—For different seasons—Summer diet—Frugivorous nations—Fresh fruit—Hygienic value of—Frugality—Its original meaning—The regimen of the Cyropædia—Roman dinners—Cornaro—Dietetic reformers—Lacour, the 'longshoreman—"*quatorze oignons*"—Commissariat of Turkish soldiers—Carlists—Experience of a newspaper correspondent—Bread as an exclusive diet—Silvio Pellico—The prisoners of the Khedive—Russian convicts—The story of Nebuchadnezzar—Polenta-meal—Diet of a Corsican farmer—Marvels of abstinence—The soldiers of Lucknow—Shamyl—Grecian philosophers—Lycurgus—Spartan soup—Philosophic symposia—The Mosaic health-code—Strange prejudices—Dietetic predilections—Papuan tidbits—Locust-eaters—*Pande-monte*—Clay-eaters—Cannibalism—The Borneo Dyaks—An Algerian anthropophage—Dr. Alcott's suggestion—Flesh-food—Popular fallacies—Cold climate—Vegetable fat *vs.* meat—Vegetarian athletes—Vegetable diet as a brain-food—Illustrations—Animal diet—Its physical and mental influence—Man a frugivorous animal—Stimulants—The poison-habit—Stimulants and poisons as synonymous terms—Unnatural habits—The secret of

CONTENTS.

their persistence—Progressive vices—Self-deceptions—Tea and coffee—Alcohol—Pungent spices—Salt—Non-stimulating diet—Its indirect advantages—Gluttony—Its principal causes—Indirect remedies—The grape-cure—Cold water in summer-time—Superstitious dread of—Occasional surfeits—Out-door exercise the best remedy—Want of appetite—Animal instincts—Dietetic caprices—Morbid appetencies—Idiosyncrasies—Dyspepsia—Curable by a change of diet—Bill of fare—General rules—Dietetic aphorisms—Intervals between meals—Four meals a day—Rest after repletion—Hygienic importance of—Jules Virey's experiments—Best time for dinner—Edinburgh tradesmen—Siesta—Fasting—Gonaque Hottentots—Eating between meals—Early habits—Dieting for health—Antiseptic foods—Hygienic conscience—General reflections.................................. 27

CHAPTER II.

IN-DOOR LIFE.

Domestic habits—Baneful prejudices—The air-famine of large cities—Saltaire—Suburban homes—Ventilation—Impure air—Simple remedies—Ventilatory contrivances—"Rain-shutters"—Bedrooms—The night-air superstition—Its utter absurdity—Illustrated by analogies—By the habits of instinct-guided creatures—By the experience of hunters and soldiers—"Draughts"—The troglodyte habit—Gaseous food—Dr. Langenbeck's experiments—Dio Lewis—Cold air not the cause of pulmonary disorders—Cave-dwellers—Opinion of a menagerie-doctor—Influence of in-door life—Suggestive statistics—Factory-slaves—The genesis of consumption—Protests of Nature—Jean Paul—City children—Their hunger after life-air—Shamyl—Popular prejudices—Ignorance not the principal cause—Mistrust in our natural instincts—Healthy homes—A model nursery—The foundling-ward of the Ursulines—Kindergarten—Nursery abuses—Domestic gymnastics—Influence of sedentary habits—Study-hours—City schools—Defective ventilation—Recess-rooms—The best writing-desk—Domestic sanitaria—"Colds"—Fallacy regarding their origin—Their real cause—Fireside comforts—Aids to domestic habits—Domestic happiness—Pets—Moral household remedies—Baths—Hydropathy—The cold-water mania—Sponge-baths—Bedrooms—The best bed—The evening-hour.. 74

CHAPTER III.

OUT-DOOR LIFE.

Out-door exercise a panacea—Its remedial influence—On the effects of intemperance—On pulmonary disorders—Hunters and herders—Their immunity from lung-diseases—The Gauchos—In-door life can become a second nature—The hygienic instincts of children—Their love of out-door sports—The paradise of childhood—Kindergarten—Bathing in sunshine—*Solaria*—Warm sand—Play-grounds—Out-door exercise—Boy-pens—Factory-children—Goethe on education—The Hydriotes—Garden-homes—Farm-work—Suggestions for city dwellers—Athletic sports—Storing up health—A remedy for colds—The perspiration-cure—Stinting our life-air—Jean Paul's aphorisms—Vacation-trips—Dr. Jordan's plan—The Ilefeld pedagogium—Tourist's outfit—Roughing it—Foot-sacks and portable beds—Benefit of pedestrian tours—In hardening the constitution—In dispelling prejudices—Wet clothes—Night-air—Cold spring-water—South-Sea Islanders—Outfit of a Canadian **lumberman—Summer** life—Refrigerating diet—Greasy made-dishes—The **cause of sun-strokes**—Camping **in** open **air—Hardiness—Modern** effeminacy—Strength not incompatible with delicacy and skill—Leonardo da Vinci—Free Saturdays—An hygienic Sabbath—Mountain excursions—Forest-air—The Puritanical Sabbath—Out-door devotion—Worshiping God in his own temple—Out-door exercise the best safeguard against certain vices—A Grecian allegory—Winter sports.. 103

CHAPTER IV.

GYMNASTICS.

Physical vigor the basis of health—Exercise **as a** peptic stimulant—Dr. Boerhaave's opinion—Curing diseases mechanically—The secret of Asklepiades—Practical proofs—A remedy *ad principium*—The interdependence of physical and moral health—Sinning against our bodies—Death-bed repentances—St. Francis of Assisi—The fallacies of asceticism—Exercise as a cosmetic—Grecian gymnastics—The neglect of physical education—Its primary cause—The war against Nature—Buddhism and the Christian ascetics—The night of the middle ages—Consequences of anti-naturalism—Physical degeneration—The lost races of Southern Europe—Revival of naturalism—Gymnastics **as a** branch of public education—Ancient gymnasia—The Olympic **games**—Out-door sports and domestic gymnastics—Children **not** naturally sluggish—Nursery exercises—Defective development—Half-made men—Undeveloped muscles—Arm-exercises—Grapple-swings and health-lifts—Precautions—Race-courses—Foot-racing—The heroes of the Iliad—Scandinavian sagas—Mexican runners—Hemerodromes—Systematic training—The drill-masters of the Janizaries—Leaping—A simple apparatus—Champion jumpers—Joe Ireland—The champion of Crotona—A daring feat—Spear-throwing—German *ger-werfers*—Target-practice—Influence of gymnastics in imparting an easy deportment—Goethe's aphorism—Climbing trees—Value of gymnastic accomplishments—Gymnastic apparatus—Asthma-cure—Lifting and carrying weights—Dr. Winship's exploits—Milo of Crotona—Wrestling—National pastimes—Revival of the Olympic festivals—A suggestion—Games of skill—Riding and swimming—Aquatic sports—Dr. Anderson's theory—Testimony of a training-ship surgeon—Analogies—Hard work—Moral influence of physical training—Boat-racing—Best time for gymnastic exercises—Female education—Physical vigor of savages—Of anthropoid apes—Of ancient and mediæval athletes—Regeneration.. 116

CHAPTER V.

CLOTHING.

Artificial teguments—Requirements of a frigid climate—The plan of Nature—Superfluous clothing—Impediments to physical development—Swaddling-clothes and flounces—Infant mortality—Statistics of Northern cities—Best dress for babies—Cold air not the cause of "colds"—Dr. Franklin's opinion—Bacon's maxim—Suggestive facts—Hardiness of savages—Of our forefathers—The dress of a Roman peasant—Darwin's Firelanders—Coale's hints on health—Tonic influence of fresh air—Air-baths—Hardy habits—Their invigorating influence on the digestive system—Disadvantages of heavy garments—Brush-coats—Triple blouses *vs.* fur—The winter dress of a fashionable lady—Dress reform—Jenny Lind—Corsets and crinolines—Warmed writing-desks—Chevalier Edelkrantz on fur mantles—Frigid climates—Winter dress of our Northern Indians—Mackinaw hoods—Neckwear—Our natural hair the best protection—Tippets and fur caps—Under-clothing—Flannel undershirts—Chamois-leather—Scotch plaids—Summer dress—Midsummer misery—The gamins of Capo-Liddo—Absurdities of modern civilization—The dread of nudity—Head-gear—Traveling bareheaded—The Emperor Hadrian—Sir John Sinclair—Adair's medical cautions—The fashion-mania—Flounces and gewgaws—The "organ of ornamentativeness"—Military uniforms—Thorwaldsen's dictum—Foot-notes—Dio Lewis's plan—Barefoot boys—The author of "Emile"—Natural sole-leather—The philosophy of clothing... 149

CHAPTER VI.

SLEEP.

Automatic functions—Sleep a restorative process—Requisite amount of sleep—For children—For adults—In disease—Dr. Page's plan—Healthfulness of village children—Due to sounder sleep—Dormitories—Silence and subdued light the best hypnotics—Abuse of narcotics—Cradling—Stupefaction not slumber—Morbid sleeplessness—Different causes of—Best remedies—Midnight serenades—A hint to nurses—The power of habit—Captain Barclay's experience—Going to sleep at short notice—Sleeping in day-time—Italian nights—The night-air dread and other superstitions—The curse of pessimism—Noontide naps—The siesta-hour—Analogies—Semi-nocturnal animals—Rambles by moonlight—Workshop misery—Sleep in disease—Nature's panacea—Pestalozzi's opinion—Insufficient sleep—Effects of—

CONTENTS.

Illustrated by historic facts—Insomnia and madness—Cure for sleeplessness—Dr. Caldwell's experiments—Robert Burton's recipe—Prognostics of longevity—Sound sleepers—Goethe and Mirabeau—Dreams—Influenced by diet—Dream-land adventures—Hard beds—Bedrooms in summer—Best location for dormitories.. 168

CHAPTER VII.

RECREATION.

Happiness our normal condition—Pessimism a symptom of disease—Periodic recreations—Hygienic importance of the festivals of the ancients—Saturnalia—Public amusements in ancient Rome—*Circenses*—Free pleasure-resorts—Asceticism—Hardships of modern civilization—Religious pessimism—Joyless laws—The "worship of sorrow"—Mirth as a remedial agent—Analogies—Recreation a physiological necessity—Ennui—Starved souls—Mirth and longevity—Dr. Brehm's experiment—Physiological influence of grief and despair—Historic examples—Inexpensive amusements—Nursery-pastimes—Playthings—Recess-hours and evening recreations—Saturday night—Vacations—A *Kinderpark*—Best recreation for school-children—Playgrounds—Woodland excursions—Fun—Happy homes—Sunday amusements—Puritanical tyranny...................................... 181

CHAPTER VIII.

REMEDIAL EDUCATION.

The health laws of Nature—Non-medicinal remedies—Prevention of disease—Medical fallacies—Suppression of symptoms—Drugs—Their legitimate use—Popular delusions—Removing the cause—Dr. Jennings's plan—Curable and incurable diseases—Definition of disease—Sickness an abnormal condition—Remedial instincts—Infantine diseases—Overfed babies—Dr. Page on nursery management—Consequences of overfeeding—The abuse of drugs—Paregoric—Scrofula—Favored by which circumstances—Dietetic remedies—Fresh air—Mediæval superstitions—The king's-evil—Radical error—Immunities of out-door laborers—Summer diet—The *Trauben-kur*—Natural remedies—Pulmonary disorders—Bronchitis—Croup—A simple remedy—Dangerous palliatives—After-effects—Worms—A symptom rather than a cause of disease—Drastic medicines—Diarrhœa—Causes of—Overeating—Irritating ingesta—Best remedy—Constipation—Aperient medicines—Objection to—Abuse of patent laxatives—The bran-bread cure—Fruit and molasses—Legumina—Graham—Out-door exercise *vs.* drugs—Secret sins—Physiological causes—Vice and indolence—Vice-centers—Gymnastics—The moral influence of physical exercise—Rickets—Mal-nutrition—Precocity—Intemperance and gluttony—Household remedies—The poison-habit—Hereditary influences—Counteracting such tendencies—Chlorosis or green-sickness—A malignant form of dyspepsia—Horseback-exercise—Tonic drugs only palliatives—The stimulant delusion—Consumption—Not an incurable disease—The records of the dissecting-room—Pulmonary consumption not necessarily hereditary—Goethe and Chateaubriand—Imprudent habits—Impure air the original cause of tuberculosis—Pulmonary scrofula—First symptoms—A crucial test—The air-cure—Forlorn hope—Pierre's case—Health without drugs—Dr. Schrodt's conclusion—Anti-naturalism—The delusions of pessimism—Analogy between moral and physical fallacies—Practical illustrations—Homœopathy—Mistrusting our natural instincts—Medical reform.................................. 198

CHAPTER IX.

HYGIENIC PRECAUTIONS.

Providential safeguards—The children of Nature—Their salutary instincts—Life under abnormal conditions—The limits of natural instincts—Intuition at fault—Mineral poisons—Curious facts—Poisons in disguise—Scutari sherbet—The cook of the Frères Provençeaux—Early impressions—Forestalling temptation—The magic of associated ideas—Fresh air—Hardiness of young children—Physiological habits—Dietetic instinct—Incidental advantages of vegetarianism—Dangers of flesh-food—Sausage-makers and their secrets—"Bologna cows"—Eggs and milk *vs.* meat—*Principiis obsta*—Nursery reform—The Laconic method—Drunken Helots—Aggressive virtue—City boys—Tobacco-smokers—Unexpected result of a physiological experiment—*Stinkewitz*—Frederick the Great—Coffee, tea, and pungent spices—Evening dinners—Dental hygiene—Cutting a third set of teeth—A natural dentifrice—St. John's bread—The diseases of the eye—Ophthalmia—Precautions—Forest-hues—Forestalling accidents—Gymnastic training the best safeguard—"Constructiveness"—Precocious prurience—Professor Weber's advice—Dietetic remedies—Hygienic rules and aphorisms. 226

CHAPTER X.

POPULAR FALLACIES.

A logical distinction—"Presumptive fallacies"—Reasoning from false premises—The natural-depravity dogma the root of the worst delusions—The leading-strings fallacy—Swaddling and cradling—A lesson from Nature—Indian babies—The children of the poor—Young South-Sea Islanders—The nostrum fallacy—The secret cause—Non-medicinal remedies—The stimulant fallacy—Dram-drinking—"Exhilarating beverages"—Tonic bitters—Self-deception—Poisons in disguise—The cold-air fallacy—Cold a popular explanation for all possible disorders—Strange delusions—"Draughts"—Catarrh—Its supposed and its real cause—The fever fallacy—Popular prejudices—Frugivorous nations **exempt from** climatic diseases—Historical facts—The origin **of fevers—Dietetic** febrifuges—The Spa fallacy—Watering-place superstition—An expensive delusion—The power of faith—Woodland **air** and mountain rambles *vs.* **mineral** water—The *Trauben-kur*—Hygienic homes—The **ascetic** fallacy—**The origin** of asceticism—Self-torturers—Supposed antagonism **of body** and soul—Joy-haters—The struggle against Nature—Its practical consequences—Perversion of natural instincts—The religion of the future—Regenesis........................ 242

PHYSICAL EDUCATION.

CHAPTER I.
DIET.

"Blessed are the pure, for they can follow their inclinations with impunity."

UNNATURAL food is the principal cause of human degeneration. It is the oldest vice. If we reflect upon the number of ruinous dietetic abuses, and their immemorial tyranny over the larger part of the human race, we are tempted to eschew all symbolical interpretations of the paradise legend, and to ascribe the fall of mankind literally and exclusively to the eating of forbidden food. From century to century the same cause has multiplied the sum of our earthly ills. Substances which Nature never intended for the food of man have come to form a principal part of our diet; caustic spices torture our digestive organs; we ransack every clime for noxious weeds and intoxicating fluids; from twenty to thirty-five per cent of our breadstuffs are yearly wasted on the distillation of a life-consuming fire; vegetable poisons, inorganic poisons, and all kinds of indigestible compounds enslave our appetites, and among the Caucasian nations of the present age an unexampled concurrence of causes has made a passive submission to that slavery the habitual condition.

Dietetic abuses, alone, would amply account for all our "ailments and pains, in form, variety, and degree beyond description"; the vitality of the human race would, indeed, have long succumbed to their combined influence, if their effects were not counteracted by the reconstructive tendency of Nature. Every birth is an hygienic regeneration. The constitutional defects which degenerate parents transmit to their offspring are modified by the inalienable bequest of an elder world—the redeeming instincts which our All-mother grants to every new child of earth. Individuals may deprave these instincts till their functions are entirely usurped by the cravings of a vicious appetency, but this perversion is never hereditary; Nature has ordained that all her children should begin the pilgrimage of life from beyond the point where the roads of misery and happiness diverge. As the golden age, the happy childhood of the human race returns in the morning of every life, the normal type of our primogenitor asserts itself athwart the morbid influences of all intermediate generations; the regenesis of every new birth brings mankind back from vice to innocence, from mysticism to realism, from ghost-land to earth. For a time those better instincts thwart the influence of miseducation as persistently as confirmed vices afterward thwart the success of reformatory measures; but, if the work of correct physical culture were begun in time, our innate propensities themselves would conspire to further its purposes and bar the boundary between virtue and vice which conscience often guards in vain. The temptations that beset the path of the adult convert, do not exist for the wards of Nature. To the palate of a normal child, alcohol is as unattractive as corrosive sublimate; the enforced inactivity of our limbs, which afterward becomes dyspeptic indolence, is as irksome to a healthy boy as to a wild animal, and a young Indian would prefer the open air of the stormiest winter night to the hot miasma of our tenement-houses. Few smokers can forget the effects of the diffident first attempt—the revolt of the system against the incipience of a virulent habit. The same with other abuses of our domestic and social life. If we would preserve the purity of our physical conscience, we might refer all hygienic problems to an unerring oracle of Nature.

The appearance of the eye-teeth (cuspids) and lesser molars marks the end of the second year as the period when healthy children may be gradually accustomed to semi-fluid vegetable substances. Till then, milk should form their only sustenance. As a substitute for the nourishment of their mother's breast, cow's-milk, mixed with a little water and sugar, is far superior to all patent paps, Liebig's compounds, and baby-soups, which often induce a malignant attack of the dysenteric complaint known as "bowel-fever" or "weaning-brash," unless palliated by still more condemnable astringents and soothing-sirups. In France the professional wet-nurses of the Pays de Vaud are generally engaged as *nourrices de deux ans;* but mothers whose employment does not interfere with their inclination in this respect may safely nurse their children for a much longer period. The wives of the sturdy Argyll peasants rarely wean a bairn before its claim is disputed by the next youngster; and the stoutest urchin of five years I ever saw was the son of a poor Servian widow, who still took him to her breast like a baby. Animals suckle their young till they are able to digest the unmodified solid food of the species; and the best method with weanlings, therefore, is perhaps that of the Ionian-

Islanders, whose toddling infants, as Dr. Bodenstedt noticed, partake of the simple repast of their parents— unleavened maize-cakes and dried figs—and are often permitted to exercise their teeth on a fresh-plucked ear of sugar-corn. But, in countries where the repast of parents is anything but simple, the best food for young children is a porridge of milk and boiled rice or oatmeal, with a little sugar, perhaps, or a few spoonfuls of apple-butter in summer-time. Of such simple dishes a child may be permitted to eat its fill, but they should be served at regular intervals and never be taken hot. Heating our food is one of the many devices for disguising its natural taste, and sipping hot and cold drinks, turn about, is far more injurious to the teeth than the *penchant* for sweetmeats which children share with savages and monkeys. Beginning with five light meals a day, the number may be gradually reduced to three, after which a system of fixed hours should be strictly observed, till the symptoms of appetite manifest a corresponding periodicity, thus saving mothers the trouble of providing baby-titbits at all possible and impossible hours of the day. Healthy children of five take readily to an exclusively vegetable diet, which is often preferable to city milk and always to flesh-food. Xenophon, in his miscellaneous "Anabasis," mentions a tribe of Bithynian coast-dwellers whose children were prodigies of chubbedness, "as thick as they were long," and remarks that said chubs were fed on—boiled chestnuts. Baked apples, pulse, macaroni, whipped eggs, bread-pudding seasoned with sugar and a drop or two of lemon-flavor, and such fruits as mellow pears, raspberries, and strawberries, can be readily assimilated by all but the weakliest nursery cadets.

But toward the end of the seventh year the advent

of a second and sturdier set of teeth suggests the propriety of exercising the jaws on more solid substances. A child of seven should graduate to a seat at the family table; or, rather, the family table should offer nothing that a child of seven can not digest. It does, though, as a rule, and parents who buy their meals ready made, or who have resigned themselves to evils from which they would save their children, should still regulate their bill of fare, both in quality and in quantity, by the rules of hygiene rather than by those of etiquette or convenience, till the age of confirmed habits puts them beyond the danger of temptation.

Before entering upon these points, I must premise a few words on the main question, What is the natural food of man? As an abstract truth, the maxim * of the physiologist Haller is absolutely unimpeachable: "Our proper nutriment should consist of vegetable and semi-animal substances which can be eaten with relish before their natural taste has been disguised by artificial preparation." For even the most approved modes of grinding, bolting, leavening, cooking, spicing, heating,†

---

* Indorsed (indirectly) in the writings of Drs. Alcott, Claude Bernard, Schlemmer, Hall, and Dio Lewis, and directly by Schrodt and Jules Virey.

† Mr. Oliphant, in his memoirs of Lord Elgin's "Mission to China," tells us that the high-caste mandarins eat all their food smoking-hot, and eschew even cold water and cold cocoa-milk, on the ground that "monkeys are addicted to such practices, which *therefore* must be injurious to a human being."

This fallacy, I believe, offers the key to the vulgar prejudice against hygienic conclusions based upon the analogies of the animal kingdom. The anti-natural doctrines of Buddhism and Christianity have persuaded us that man is an *alter ens*, a being governed by laws opposed to those of Nature in general, and that arguments derived from the habits of our dumb fellow-creatures can not be validly applied to the problems of human physiology. "You do not compare yourself to a wild animal, do you?" is a common objection to such arguments. Wild animals have to and freezing our food are, strictly speaking, abuses of our digestive organs. It is a fallacy to suppose that hot spices aid the process of digestion: they irritate the stomach and cause it to discharge the ingesta as rapidly as possible, as it would hasten to rid itself of tartarized antimony or any other poison; but this very precipitation of the gastric functions prevents the formation of healthy chyle. There is an important difference between rapid and thorough digestion. In a similar way, a high temperature of our food facilitates deglutition, but, by dispensing with insalivation and the proper use of our teeth, we make the stomach perform the work of our jaws and salivary glands; in other words, we make our food less digestible. By bolting our flour and extracting the nutritive principle of various liquids, we fall into the opposite error: we try to assist our digestive organs by performing mechanically a part of their proper and legitimate functions. The health of the human system can not be maintained on concentrated nutriment; even the air we inhale contains azotic gases which must be separated from the life-sustaining principle by the action of our respiratory organs—not by any inorganic process. We can not breathe pure oxygen. For analogous reasons bran-flour makes better bread than bolted flour; meat and saccharine fruits are healthier than meat-extracts and pure glucose. In short, artificial extracts and compounds are, on the whole, less wholesome than the palatable products of Nature. In

rely on the guidance of their natural instincts; but those instincts teach them to avoid poisons, and to cure their diseases without drugs; they teach them not to murder their unborn offspring, not to eat till they are hungry, not to starve in the midst of plenty, not to choke for fear of the night-air, not to fuddle with alcohol or opium. We are better Christians and better pastry-cooks, but in some respects it would be well for us if we had the *right* to " compare ourselves to wild animals."

the case of bran-flour and certain fruits with a large percentage of wholly innutritious matter, chemistry fails to account for this fact, but biology suggests the mediate cause: the normal type of our physical constitution dates from a period when the digestive organs of our (frugivorous) ancestors adapted themselves to such food—a period compared with whose duration the age of grist-mills and made dishes is but of yesterday.

We can not doubt that the highest degree of health could only be attained by strict conformity to Haller's rule, i. e., by subsisting exclusively on the pure and unchanged products of Nature. In the tropics such a mode of life would not imply anything like asceticism: a meal of milk and three or four kinds of sweet nuts, fresh dates, bananas, and grapes would not clash with the still higher rule, that eating, like every other natural function, should be a pleasure and not a penance. Heat destroys the delicate flavor of many fruits, and makes others less digestible by coagulating their albumen. But in the frigid latitudes, where we have to dry and garner many vegetable products in order to survive the unproductive season, the process of cooking our food has advantages which fully outweigh such objections. Few men with post-diluvian teeth would agree with Dr. Schlemmer that hard grain is preferable to bread. No Bostoner would renounce his favorite dish for a nose-bag full of dry beans. Dried prunes, too, are improved by cooking—in taste, at least, and perhaps in digestibility. Besides, we should not forget that the natural taste of such substances, before they became over-dry, *was* agreeable, or at least not repulsive to our palates.* It appears that on week-days the chil-

---

* In his "Natur-Heilkunde," Schrodt distinguishes between natural, artificially adapted, and unnatural or wholly injurious articles of food.

dren of Israel indulged their poor in the practice of snatching free luncheons from a convenient corn-field (Matthew xii, 1), and the Imam of Muscat still feeds his soldiers on crude wheat and dhourra-corn, a sort of millet, which many French soldiers learned to eat raw, as their Mameluke captors declined to cook it for them. Even the legumes — peas, beans, and lentils — pass through a period when they are soft and full of sweet milk-juice, though in their sun-dried over-ripeness they become as tough as wood. In the scale of wholesomeness the place next to Haller's man-food *par excellence* should therefore be assigned to vegetable substances whose pleasant taste has been *restored* by the process of cooking. With this addition, even an invalid, dieting for his health, need not complain of lack of variety, for the number of nutritious vegetables that can be successfully cultivated as far north as Hamburg and Boston is almost infinite if we include the plants of the corresponding Asiatic latitudes and those that could be acclimatized in the course of five or six seasons. With five kinds of cereals, three legumina, eight species of escu-

"Our natural food," he says (like Pythagoras), "are such vegetable and semi-animal products as *either are or can be eaten and relished raw, and without the preliminaries of cooking and spicing*. Such are milk, honey, eggs, nuts, cereals, a few roots, legumina, and gums, and the countless variety of fruit, which are man-food *par excellence*. Our various kinds of bread, though artificially prepared, as well as other farinaceous dishes, are derived from an edible grain which is neither repulsive nor indigestible in its original state.

"To the second or adapted edibles belong different vegetables which are rendered palatable only by the process of cooking, as cabbage, beans, peas, and lentils, and various roots and leaves. Flesh, also, I will add to this list, though some would place it in the third class. Injurious, without a redeeming quality, are all narcotic and alcoholic drinks, and all ardent spices, such as pepper, mustard, and acid fluids; also those partly decayed and acid substances whose properties are more stimulating than nourishing: strong cheese, sauerkraut, and pickles."

lent roots, ten or twelve nutritive herbs, thirty to forty varieties of tree-fruits, besides berries and nuts, a vegetarian might emulate the Duc de Polignac, who refused to eat the same dish more than once per season. Honey is the pure, unchanged, **and unalloyed** saccharine juice of flowers and resinous exudations, and therefore strictly a **vegetable substance**, though Carl **Bock** and Bichat describe it as semi-animal food, because "derived from animals," i. e., hived by bees. They might as well include flour under the same category because horses carry grist to the mill. Like sugar, vanilla, and the manna-sirup of Arabia Felix, we might class it with the non-stimulating condiments, which, used in moderate quantities, impart an agreeable flavor to many farinaceous preparations without impairing their digestibility.*

Of all semi-animal substances, sweet fresh milk is the most wholesome, in itself an almost perfect aliment, welcome to all mammals and nearly all vertebrate animals. Monkeys, cats, deer, squirrels, otters, and ant-bears, creatures that differ so widely in their *special* diet, will rarely refuse a dish of this **universal** food. I have seen snakes and iguanas drink it with avidity. On the other hand, I have noticed that all animals but pigs and starved dogs eschew sour milk; it is, properly speaking, fermented **milk**, to the taste of a normal man probably as repulsive as tainted meat or sour gruel. This fermentation affects the fatty particles less than the watery and caseine; and butter and cream (though less digestible than fresh milk) are, therefore, far

---

* Jules Virey estimates that four tenths of the human race subsist exclusively on a vegetable diet, and that seven tenths are practically (though not on principle) vegetarians. Virchow estimates the total number at eighty-five per cent.

healthier than sour whey and cheese. Cheese in some of its forms is quite as unwholesome as rotten flesh; putrid curd would be the right name for Limburger and fromage de Brix. Vegetarians of the Lankester school object to milk and butter on account of the spurious stuff that is often foisted upon the market under those names; but mild-tasted aliment scan hardly be adulterated with very injurious substances; a little tallow, oleomargarine, or even lard, mixed with butter, and as such again mixed with a tenfold quantity of farinaceous food, can only affect the most delicate constitutions to any appreciable degree, and certainly not more than the small percentage of alum we often eat with our daily bread. Comparatively speaking, such things are the veriest trifles, and we can not afford to fight gnats while we are beset by a swarm of vampires. We have dietetic exquisites who would shudder at the idea of raising their biscuits with brewer's yeast instead of bicarbonate of soda, but do not hesitate to sandwich that same bread with strong cheese and pork-sausage; or pity the wretch whose poverty consents to North Carolina apple-jack, while they sip a *petite verre* of aromatic schiedam. That kind of purism often reminds me of the fastidiousness of Heinrich Heine's Mandarin convict, who insists on being thrashed with a perfumed bamboo, "but would have been shocked at a less fragrant hiding."

All kinds of fat ("non-nitrogenous" aliments), including butter and cream, are more digestible in winter than in summer time. Cold air is a peptic stimulant, and neutralizes the calorific effect of a non-nitrogenous diet, while fresh tree-fruits and berries counteract an excess of atmospheric heat, and thus, by an admirable provision of Nature, the seasons themselves furnish us the food most adapted to the preservation of the right medium temperature of the system. Preserved fruits (raisins, dried figs and apples, etc.) lose much of their acidity, and thus become less refreshing, but not less nutritive, at the very time when the latter property is the more important one. Cow's-milk, on the other hand, grows richer in winter-time, and this self-adaptation of their food to the varying demands of the seasons enables the inhabitants of such countries as Italy and Mexico to subsist all the year round on an almost uniform diet. But in a climate of such thermal extremes as ours it would be the best plan to vary our regimen with the weather, and, above all, to adopt a special summer diet, since the consequences of our present culinary abuses are far less baneful in January than in July. Even in mid-winter our compounds of steaming and greasy viands with hot spices severely strain the tolerance of a youthful stomach; but, when the dog-star adds its fervid influence, the demand for refrigerating food becomes so imperative that no forensic eloquence would persuade me to convict a city lad for hooking water-melons. Where fruit is cheap, the *paterfamilias* should keep a store-room full of summer apples, and leave the key in the door—it will obviate costiveness and midnight excursions. From May to September fresh fruit ought to form the staple of our diet, and the noonday meal at least should consist of cold dishes, cold apple-pudding with sweet milk and whipped eggs, or strawberries with bread, cream, and sugar. The Romans of the republican age broke their fast with a biscuit and a fig or two, and took their principal meal in the cool of the evening. In their application of the word, a frugal diet meant quite literally a diet of tree-fruits, and that our primogenitor was a frugivorous creature is the one point in which the Darwinian genesis agrees with the Mosaic version.

Cyrus, King of Persia, according to Xenophon, was brought up on a diet of water, bread, and cresses, till up to his fifteenth year, when honey and raisins were added; and the family names of the Fabii and Lentuli were derived from their customary and possibly exclusive diet. Eggs and apples, with a little bread, were for centuries the alpha and omega of a Roman dinner; and, in earlier times, even bread and turnips, if not turnips alone, which the patriot Cincinnatus thought sufficient for his wants. It is singular that our temperance societies direct their efforts only against the fluid part of our vicious diet; a league of temperate eaters would certainly find a large field for reform. But in Italy the thing was attempted by Luigi de Cornaro, a Venetian nobleman of the fifteenth century, who restricted himself to a daily allowance of ten ounces of solid food and six ounces of wine, and prolonged his life to one hundred and two years. Though he did not organize his followers into a sect, his example and his voluminous writings influenced the manners of his country for many years. Cornaro would not have gained many converts in Russia and Germany; but throughout Southern Europe frugality, in the truest old Latin sense, is by no means rare. Lacour, a Marseilles 'longshoreman, earned from ten to twenty francs a day, loaned money on interest and gave alms, but slept at night in his basket, and subsisted on fourteen onions a day, which preserved him in excellent health and humor, but got him the nickname of *quatorze oignons.*

A pound of bread with six ounces of poor cheese, and such berries as the road-side may offer, constitute the daily ration of the Turkish soldier on the march, and the followers of Don Carlos contented themselves with even less. A correspondent of the "Daily News" was served with a dish of radishes in a Catalan tavern, and ventured the remark that radishes were taken after meals in Northern Europe. "You can get some more after finishing these," was the reply. The radishes constituted the dinner.

Not that men *should*, but that they *can*, live on bread alone, is abundantly proved by the records of Old-World prisons. Silvio Pellico, the Italian patriot and martyr, subsisted for seven years on coarse rye-bread and water, which experience had taught him to prefer to the putrid pork-soup of his Austrian bastile. The prisoners of the Khedive were fed on rice and Indian corn, till the prayers of the French residents and his American officers induced him to sweeten their bitter lot by a weekly bottle of sakarra, or diluted molasses; and I learn from an article in a French journal that some of these unfortunates, who had passed long years without any hint of sakarra, were forced by chronic bowel complaints to return to their old dry fare.

Fedor Darapski, born 1774 in Karskod near Praga, Eastern Poland, was brought to the government of Novgorod in his twenty-second year as a conscript to the Russian army, and was soon after sentenced to death for mutiny and assault with intent to kill. The Empress Catharine, acting on a recommendation of the Governor of Novgorod, commuted his sentence to imprisonment for life, but ordered that on every anniversary of the deed (an attempt to kill his colonel) the convict should receive forty lashes and be kept on half rations for a week after; the full ration being two pounds of black bread and a jug of cold water. On these terms Darapski was boarded at the fortress of

Kirilov till 1863, when at the approach of his ninetieth birthday he was again recommended to mercy, and liberated by order of the late Czar.

Even the story of Nebuchadnezzar may be more than an allegory, as the wild berries, roots, and grass-seeds of the Assyrian valleys contained surely as much nourishment as sour rye-bread; and who knows but grass itself might do for a while, since the Slavonian peasants often subsist for weeks at a time on sauerkraut and cabbage-soup?

Corsican farmers live all winter on dried fruit and *polenta* (chestnut-meal), and the Moors of mediæval Spain used to provision their fortified cities with chestnuts and olive-oil. During the siege of Lucknow the native soldiers asked that the little rice left be given to their British comrades; as for themselves, they could do with the *soup*, i. e., the water in which the rice had been boiled!

But the *ne plus ultra* of abstinence combined with robust strength is furnished in the record of Shamyl, the heroic Circassian, who for the last two years of the war that ended with his capture had nothing but water for his drink and roasted beechnuts for his food, and yet month after month defied the power of the Russian Empire in his native mountains, and repeatedly cut his way through the ranks of his would-be captors with the arm of a Hercules.

The philosophers of antiquity prided themselves on their frugal habits, which ranked next to godliness in their estimation, as expressed in the famous aphorism, "God needs nothing, and he is next to him who can do with next to nothing"—whose material needs are the smallest. Primitive habits are certainly favorable to independence, especially in a genial climate, where a man is above the fear of tyranny and all social obligations, who like Shamyl can subsist on the spontaneous gifts of his mother Earth. "Do you know," Cyrus asked the embassador of a luxurious potentate, "how invincible men are who can live on herbs and acorns?" If the Saracens had persisted in the simplicity of their fathers, the nineteenth century might see Moorish kingdoms in Southern Europe, and Arabian science and fruit-gardens in the place of deserts and monkish besottedness. Cato needed no prophetic inspiration to predict the downfall of a city where a small fish could fetch a higher price than a fattened ox.

Lycurgus, the Spartan, makes the diet of his countrymen the subject of careful legislation, but seems to have feared excesses in quality rather than in quantity; as long as the black soup and other national dishes remained orthodox in regard to the prescribed simple ingredients, free indulgence of the most exacting appetites was not only permitted but encouraged. At the philosophic reunions of the Lyceum the bill of fare permitted a choice between dried figs and honey-water in addition to the wheat-bread, which could not be refused, and Greece was the model of early Roman institutions in this as well as in other respects. Fruit and bread-cakes, spiced with Attic salt and music, entertained the friends of Plato at those suppers of the gods of three or four hours, which Aristotle preferred to so many years on the throne of Persia; but the very next generation witnessed the drunken riots of Babylon, and the general introduction of Persian manners and luxuries.

The ancients undoubtedly were our superiors in hygienic insight, but among the many judicious restrictions of their dietary regimens there are some that we must attribute to **prejudice or leave** utterly unaccounted for. The Mosaic interdiction of rabbit-flesh, wild swan, and finless fishes has been very learnedly explained as a necessary consequence of general laws, which had to include those animals for the sake of consistency; but what on earth or below earth could induce Pythagoras, the great philosopher, to prohibit the use of *beans* —nay, even denounce any contact with the shell, the leaves, or the roots of the poor plant as a dreadful pollution? Such was the stigma he had attached to the violation of this rule, we are told, that a body of soldiers from Magna Græcia, who all belonged to the Pythagorean sect, permitted themselves to be cut to pieces or captured rather than save themselves by crossing a bean-field!

The old proverb *de gustibus* can hardly prevent astonishment at the diversity of tastes. What would Pythagoras have said about our national dish of pork and beans, or what shall we say to explain the Japanese prejudice against milk, the Papuan's partiality for fat white caterpillars, or the *gliraria* that were attached to every decent household of imperial Rome? Athenæus describes a glirarium as a large brick structure, divided by wire partitions into small cells, from five hundred to two thousand of them; every cell the receptacle of a captive rat, which was fattened on husks, rotten fish, and other offal, till a further increase in bulk would make it difficult to extract the animal through the narrow door of its cage. The perfect specimens were then collected, stuffed with crushed, figs, and served in a sauce of olive-oil at the banquets of wealthy patriots who preferred domestic delicacies to colonial imports. The Digger Indians of our Pacific slope rejoiced in the great locust-swarms of 1875 as a gracious dispensation of the Great Spirit, and laid in a store of dried locust-powder for years to come. Even mineral substances and strong mineral poisons have their votaries. Mithridates, King of Pontus, could take a large dose of arsenic with impunity, and the mountaineers of Savoy and Southern Switzerland use arsenic habitually as a safeguard against pulmonic affections. The poor Norsemen often mix their daily bread with a whitish mineral powder, more from necessity than a vitiated taste, we hope; but a similar substance is employed by the natives of Brazil and other parts of tropical America without any such excuse. The name of Panama is derived from *panamante* (originally *pan-de-monte*, mountain-bread), a substance which the Indians of Central America prepared from a mealy gypsum powder, found here and there in the Sierra. Humboldt describes a tribe of Indians in Northern Brazil who have been addicted to the use of panamante for generations, and were distinguished by a monstrous protuberance and induration of the upper abdomen. When the French were masters of St. Domingo their negro slaves had contracted a similar passion, and could only be restrained by barbarous punishments from indulging it to excess.

It would be erroneous to suppose that cannibalism has become quite extinct. Among the Dyaks of Borneo there is a recurrence of the outrage after every petty feud and raid, and many of the South-Sea Islands are still infested with secret anthropophagi. The Pintos, an aboriginal tribe of Yucatan, have repeatedly been detected in cannibal practices; and phenomenal cases have occurred in Asia after every protracted famine. In 1873 the Chasseurs d'Afrique captured an old Kabyle on the plateau of Sidi-Belbez (Algiers), who had committed innumerable murders to indulge this horrible passion, and had twice been caught *in flagrante* by his countrymen, who contented themselves with giving him a good hiding the first time, and released him on another occasion when they found his victim had only been a French settler!

The slaughter-houses of every large city are visited by delicate ladies, who hope to cure affections of the respiratory organs by a draught of fresh blood, but who would inspire a Hindoo with a cannibal terror more intense than that produced in the Algerian settlements by the above Kabyle. Herodotus relates that the Scythians executed their criminals by a potion of fresh oxblood, and recommends this as a more humane method than capital punishment by the sword, though inferior to the hemlock-cup. "For opening the gates of Tartarus," says Haller, "there is nothing like a good narcotic. If I should have occasion to leave this world, I would no more think of shooting myself than of leaving town by being fired from a mortar, when I could take the stage-coach."

The Turks shudder at seeing a Frank swallow oysters, and even in the cities of Europe and North America we find individuals with similar antipathies; and I know an old professor who passed half a century in St. Petersburg, and suffered grievously from an unconquerable aversion to caviare. Caviare is the salted or pickled roe of the sturgeon—not quite so bad as Schnepfendreck, a North German delicacy, which consists chiefly of the fæces of the common woodcock.

Professor H. Letheby, food-analyst for the city of London, is responsible for the following account of a mandarin's dinner, given to an English party and some distinguished natives of Hong-Kong:

"The dinner began with hot wine, made from rice, and sweet biscuits of buckwheat. Then followed the first course of custards, preserved rice, fruits, salted earth-worms, smoked fish and ham, Japan leather (?) and pigeons' eggs, having the shells softened by vinegar; all of which was cold. After this came sharks' fins, birds' nests, deer-sinews, and other dishes of an appetizing and dainty character. They were succeeded by more solid foods, as rice and curry, chopped bears' paws, mutton and beef cut into small cubes and floating in gravy; pork in various forms, the flesh of puppies and cats boiled in buffalo's milk; shantung or white cabbage and sweet-potatoes; fowls split open, flattened and grilled, their livers floating in hot oil, and cooked eggs of various descriptions, containing embryo birds. But the surprise of the entertainment was yet to come. On the removal of some of the flower-vases a large covered dish was placed in the center of the table, and at a signal the cover was removed. The hospitable board immediately swarmed with juvenile crabs, who made their exodus from the vessel with surprising agility, for the crablets had been thrown into vinegar before the guests sat down, and this made them sprightly in their movements; but, fast as they ran, they were quickly seized by the nearest guests, who thrust them into their mouths and crushed them without ceremony, swallowing the strange gelatinous morsel with evident gusto. After this, *soy* was handed round, which is a liquor made from a Japan bean, and is intended to revive the jaded palate. Various kinds of shell and fresh fish followed, succeeded by several thin broths. The banquet was concluded by the costly bird's-nest soup, the dessert being a variety of scorched seeds and nuts, with sundry hot wines and tea."

But the mandarin was astonished in his turn by finding ice-cream among the delicacies of an English refreshment-table, and predicted disastrous consequences from its habitual use. Ice, without doubt, is injurious, but not more unnatural than our custom of swallowing boiling-hot soups and stews.

Dr. Alcott holds that a man might live and thrive on an exclusive diet of well-selected fruits, and I agree with him if he includes olives and oily nuts, for no assumption in dietetics is more gratuitous than the idea that a frequent use of flesh-food is indispensable to the preservation of human health. Meat is certainly not our *natural* food. The structure of our teeth, our digestive apparatus, and our hands, proves *a priori* that the physical organization of man is that of a frugivorous animal. So do our instincts. Accustom a child to a diet of milk, bread, and meat; never let him see a fruit, nor mention the existence of such a thing; then take him to an orchard, and see how quickly his instinct will tell him what apples are good for. Turn him loose among a herd of lambs and kids; he will play with them as a fellow-vegetarian. In a slaughter-house the sight of gory carcasses and puddles of blood will excite him with a *horror naturalis*. The same sight would excite the *appetite* of the omnivorous pig as well as of the carnivorous puppy. Artificial preparation, spices, etc., may disguise the natural taste of meat, as of coffee or wine, but they will not alter its effect upon the animal system. The flesh-food fallacy, like other errors of the civilized nations, has found plausible defenders, but their principal argument is clearly based on a misunderstood fact. The delusion originated in England, where the *physique* of the beef-fed and rubicund Saxon squire contrasts strongly with that of the potato-fed Celtic laborer. What this really proves is merely that a mixed diet is superior to a diet of starch and water, for the North Irish dairyman, who adds milk and butter to his starch, outweighs and outlives the rubicund squire. The matter is this: in a cold climate we can not thrive without a modicum of fat, but that fat need not come from slaughtered animals. In a colder country than England, the East-Russian peasant, remarkable for his robust health and longevity, subsists on cabbage-soup, rye bread, and vegetable oils. In a colder country than England, the Gothenburg shepherds live chiefly on milk, barley, bread, and esculent roots. The strongest men of the three manliest races of the present world are non-carnivorous: the Turanian mountaineers of Daghestan and Lesghia, the Mandingo tribes of Senegambia, and the Schleswig-Holstein *Bauern*, who furnish the heaviest cuirassiers for the Prussian army and the ablest seamen for the Hamburg navy. Nor is it true that flesh is an indispensable, or even the best, brain-food. Pythagoras, Plato, Seneca, Paracelsus, Spinoza, Peter Bayle, and Shelley, were vegetarians; so were Franklin and Lord Byron in their best years. Newton, while engaged in writing his "Principia" and "Quadrature of Curves," abstained entirely from animal food, which he had found by experience to be unpropitious to severe mental application. The ablest modern physiologists incline to the same opinion. "I use animal food because I have not the opportunity to choose my diet," says Professor Welch, of Yale; "but, whenever I have abstained from it, I have found my health mentally, morally, and physically better."

Though a vegetarian on principle, I have eaten various kinds of flesh as a physiological experiment, and have often observed the influence of animal food upon children and invalids, and I have found that a pound of boiled beef or eight ounces of lean pork, after a month's abstinence from all flesh-food, will infallibly produce some or all of the following unmistakable effects: a gastric uneasiness, akin to the incipient operation of certain emetics; distressing dreams, restlessness, and a peculiar mood which I might describe as a promiscuous pessimism, a feeling of general irritation and resentment. I have also noticed that flesh-food tends to check intellectual activity, not so much by making us averse to all mental occupations as by muddling what phrenologists call the *perceptives*. By its continued use children gradually lose their native brightness as well as their amiable temper.

But the same observations oblige me to say that its deleterious *physical* effects have often been considerably overrated. The gastric uneasiness, even after a hearty meal of meat (fat pork, perhaps, excepted), yields readily to exercise in open air. Meat does not interfere with the digestion of other food, and, above all, it produces no ruinous after-effects; its frequent use rarely becomes a morbid necessity. Besides, flesh undoubtedly contains many nutritive elements, though in a less desirable form than we might find them in vegetable substances. By dint of practice the system can be got to accept part of its nutriment in that form, and if we are reduced to the choice of starving on starch and watery herbs, or getting fat in an abnormal way, the latter is clearly the preferable alternative. As a rule, though, children during their school years had better stick to dairy products, farinaceous preparations, and fruit; hot-headed boys, especially, can be more effectually cured with cow's-milk than with a cow-hide.

The objections to flesh-food, however, do not apply to eggs, and not in the same degree to mollusks and crustaceans. On the banks of the Essequibo, in Eastern Venezuela, I have seen troops of capuchin monkeys (*Cebus paniscus*) engaged in catching crabs, though in captivity those same relatives of ours would rather starve than touch a piece of beef. The dog-headed baboon visits the sea-shore in search of mollusks, and the South American marmoset, like John the Baptist, delights in grasshoppers and wild honey, though otherwise a strict vegetarian. The mediæval distinction between flesh and fish is not wholly gratuitous, either; carp, trout, and their congeners are, happily, almost as digestible as potatoes, for it would be a hopeless undertaking to dissuade a young Walton from boiling and devouring his first string of perch. On journeys, especially in cold weather, children may be occasionally indulged in such way-side delicacies as codfish-balls, oiled sardines, and ham-sandwiches.

But, under all circumstances, make a firm stand against the POISON-HABIT. It is best to call things by their right names. The effect upon the animal economy of every stimulant is strictly that of a poison, and every poison may become a stimulant. There is no bane in the South American swamps, no virulent compound in the North American drug-stores—chemistry knows no deadliest poison—whose gradual and persistent obtrusion on the human organism will not create an unnatural craving after a repetition of the lethal dose, a morbid appetency in every way analogous to the hankering of the toper after his favorite tipple. Swallow a tablespoonful of laudanum or a few grains of arsenious acid every night: at first your physical conscience protests by every means in its power; nausea, gripes, gastric spasms, and nervous headaches warn you again and again; the struggle of the digestive organs against the fell intruder convulses your whole system. But you continue the dose, and Nature, true to her highest law to preserve life at any price, finally adapts herself to an abnormal condition—adapts your system to the poison at whatever cost to health, strength, and happiness. Your body becomes an opium-machine, an arsenic-mill, a physiological engine moved by poison, and performing its vital functions only under the spur of the unnatural stimulus. But by-and-by the jaded system fails to respond to the spur, your strength gives way, and, alarmed at the symptoms of rapid *deliquium*, you resolve to remedy the evil by removing the cause. You try to renounce stimulation, and rely once more on the unaided strength of the *vis vitæ*. But that strength is almost exhausted. The oil that should have fed the flame of life has been wasted on a health-consuming fire. Before you can regain strength and happiness, your system must *readapt* itself to the normal condition, and the difficulty of that rearrangement will be proportioned to the degree of the present disarrangement; the further you have strayed from Nature, the longer it will take you to retrace your steps. Still, it is always the best plan to make your way back somehow or other, for, if you resign yourself to your fate, it will soon confront you with another and greater difficulty. Before long the poison-fiend will demand a larger fee; you have to increase the dose. The "delightful and exhilarating stimulant" has palled, the *quantum* has now to be doubled to pay the blue-devils off, and to the majority of their distracted victims that seems the best, because the shortest, road to peace. Restimulation really seems to alleviate the effects of the poison-habit for a time. The anguish always returns, and always with increased strength, as a fire, smothered for a moment with *fuel*, will soon break forth again with a fiercer flame.

By these symptoms the disease of the poison-habit may be identified in all its disguises, for the self-deception of the poor lady who seeks relief in a cup of the same strong tea that has caused her sick-headache is absolutely analogous to that of the pot-house sot who hopes to drown his care in the source of all his misery, or of the frenzied opium-eater who tries to exorcise a legion of fiends with the aid of Beelzebub. There are few accessible poisons which are not somewhere abused for the purpose of intoxication: the Guatemala Indians fuddle with hemlock-sap, the Peruvians with *coca*, the Tartars with fermented mare's milk, the Algerians with hasheesh; but, wherever men have dealings with the "fiend that steals away their brains," there are always Ancient Iagos who mistake him for a "good familiar creature," till he steals their health and wealth as well as their wits. Their woes are not the penalty of their persistent blindness, but of their first open-eyed transgression. There is a Spanish proverb to the effect that it is easier to keep the devil out than to turn him out, and many dupes of the Good Familiar would actually think it an ingratitude to turn him off; but they should have known better than to admit him when he presented himself with horns and claws. To a normal taste every poison is abhorrent, and with the rarest exceptions the degree of the repulsiveness is proportioned to that of the virulence. In the mouth of a healthy child, rum is a liquid fire; beer, an emetic; tea and coffee, bitter decoctions; tobacco-fumes revolt the stomach of the non-*habitué*. Only blind deference to the example of his elders will induce a boy to accustom himself to such abominations; if he were left to the guidance of his natural instincts, intoxication would be anything but an insidious vice.

With all its ramifications, the poison-habit is a upas-tree which has polluted the well-springs and tainted the very atmosphere of our social life. The woe which the human race owes to alcohol alone is so far beyond description that I will here only record my belief that its total interdiction will form the first commandment in the decalogue of the future. The power of prejudice has its limits. No man, possessed of a vestige of common-sense, can read the scientific literature that has accumulated upon the subject, and doubt that even the moderate use of distilled liquors as a beverage amply justifies the belief in the existence of unqualified evils. The effects of tea and coffee drinking are also well understood, but I must call attention to an often overlooked though most important feature of the habit—its progressiveness. The original moderate *quantum* soon palls, and it is this craving of the system *for the same degree of stimulation* which leads us to Johnsonian excesses or to the adoption of a stronger stimulant. Men generally prefer the latter alternative. Coffee, tea, and tobacco pave the way to opium in the East and to alcohol in the West. The same holds true of pungent spices. Pepper and mustard form the vanguard of the poison-fiend. They inflame the liver, produce a morbid irritability of the stomach, cause numerous functional derangements by impeding the process of assimilation, and thus become auxiliary in expediting the development of the poison-habit. Whatever irritates the digestive organs or unusually exhausts the vital forces tends to the same effect. Besides, they blunt the susceptibility of the gustatory nerves, and thus diminish our enjoyment of the simple viands that should form our daily food. In trying to heighten that enjoyment, the surfeited gastronome defeats his own purpose: all sweetmeats pall; the most appetizing dishes he values only as a foil to his caustic condiments, like the Austrian peddler who trudges through the flower-leas of the Alpenland in a cloud of nicotine, and to whom the divine afflatus of the morning wind is only so much draught for his tobacco-pipe.

With a single and not quite explained exception, man is the only animal that resorts to stimulation: a few ruminant mammals—cows, sheep, and deer—pay an occasional visit to the next salt-lick. The carnivora digest their meat without salt; our next relatives, the frugivorous four-handers, detest it. Not one of the countless tonics, cordials, stimulants, pickles, and spices, which have become household necessities of modern civilization, is ever touched by animals in a state of nature. A famished wolf would shrink from a "deviled gizzard." To children and frugivorous animals our pickles and pepper-sauces are, on the whole, more offensive than meat, and therefore probably more injurious. To savages, too. In the summer of 1875 I stood one evening near the quartermaster's office at Fort Wingate, New Mexico, when two Kiowa Indians applied for permission to water their famished horses at the government cistern, offering to accept that boon in part payment of a load of brushwood which they proposed to haul from the neighboring *chaparral*. The fellows looked thirsty and hungry themselves, and, while the quartermaster ratified the wood-bargain, one of the officers sent to his company quarters for a lunch of such comestibles as the cooks might have on hand at that time of the day. A trayful of "government grub" was deposited on the adjacent cord-wood platform, and the Indios pitched in with the peculiar appetite of carnivorous nomads. A yard of commissary sausage was accepted as a tough variety of jerked beef; yeasted and branless bread disappeared in quantities that would have confirmed Dr. Graham's belief in natural depravity; they sipped the cold coffee and eyed it with a gleam of suspicion, but were reconciled by the discovery of the saccharine sediment, and the cook was just going to replenish their cups when the senior Kiowa helped himself to a vinegar pickle, which he probably mistook for some sort of an off-color sugar-plum. He tasted it, rose to his feet, and dashed the plate down with a muttered execration, and then clutched the prop of the platform to master his rising fury. Explanations followed, and a pound of brown sugar was accepted as a peace-offering; but the children of Nature left the post under the impression that they had been the victims of a heartless practical joke. "D—n their breechless souls, they don't know what's good for them!" was the cook's comment, which I should indorse if his guests had been in need of a blister. A slice of peppered and allspiced vinegar pickle will blister your skin as quick as a plaster of Spanish flies. The lady-friends of Dio Lewis have promised us an "Art of Cookery for Total Abstainers," and, if the book should correspond to the title, I would suggest a motto: "No spice but hunger; no stimulant but exercise."

In the use of hot spices the Spaniards and their South American kinsmen exceed every other nation. *Chilé colorado*, or red pepper, is one of the mildest condiments of a Peruvian kitchen. The *yerba blanca*, a whitish-green herb which is used raw with olive-oil on sandwiches, and enters into the composition of various ragouts, is described as resembling the *lapis infernalis* in its effect on a normal tongue. A Mexican can chew up a handful of red pepper as we would so much dried fruit, and eats onions, garlic, and salted radishes as a relief from more pungent tastes. I must believe it, on the testimony of the entire medical faculty of the city of Bremen, that a man who was treated in their city hospital for a most mysterious complaint settled the dispute of his physicians by confessing a weakness for *tan-water*—the fiery infusion of tan-bark, in which he had indulged rather to excess in the last year. The inhabitants of Southern Russia, especially of the Dnieper Delta, are all day long chewing the aromatic seeds of the sunflower and different kinds of pumpkin-seeds, which appears to be less a stimulation than an idle habit, like the use of chewing-gum in our boarding-schools.

Timour the Tartar celebrated his victories by solemn barbecues of broiled horse-flesh and fermented mare's milk, or koumiss, which is still a favorite drink of his countrymen. Tartars also use a decoction of the poisonous fly-sponge as a stimulating beverage, and according to Vambéry have a national foible for morsels of superannuated meat, of an aroma which the French term of *haut-goût* would hardly begin to describe. Yet these same Tartars might shudder at being confronted with a dish of that Limburg delicacy which finds its way into the best hotels of continental Europe. I cannot forget the emphatic protest of a Spanish officer who was invited to partake by a German admirer of the questionable dainty, in the cabin of a Havana steamer. "You think it unhealthy to eat that?" inquired the Hamburger, in polite astonishment. "Unhealthy?" exclaimed the Hidalgo, with a withering look and a gasp for a more adequate word—"no, sir! I think it an unnatural crime!"

Assassin, assassinate, and their derivatives, come from *hasheesh,* the Arabian word for hemp. A decoction of hemp-leaves, filtered and boiled down, yields a greenish-black residuum of intensely bitter and nauseous taste—a stuff not very likely, one should think, to tempt a normally constituted human being. Yet this same hasheesh, Dr. Nachtigal assures us, can marshal a larger army of victims than either gunpowder or alcohol; and only the originator of the opium-habit, he thinks, will have an uglier score against him on the day of judgment than the Sheik-al-Jebel, who, tradition says, first introduced the hasheesh-habit. A frugal diet has this additional advantage, that simple food is in less danger of adulteration, or must at least be imitated by equally simple and harmless substitutes. Watered milk or lard mixed with corn-meal is certainly annoying, but hardly injurious, and is a trifle altogether if compared with the abominations that are half consciously consumed by the lovers of imported delicacies and expensive stimulants. Dr. Stenhouse, of Liverpool, analyzed a suspicious sample of tea, with the following result, published in the "Planters' Price Current" of February, 1871: The package contained some pure congou-tea leaves, also siftings of pekoe and inferior kinds, weighing together twenty-seven per cent of the whole. The remaining seventy-three per cent were composed of the following adulterants: Iron, plumbago, chalk, china-clay, sand, prussian-blue, turmeric, indigo, starch, gypsum, catechu, gum, the leaves of the camellia, sarangua, *Chlorantes officinalis,* elm, oak, willow, poplar, elder, beech, hawthorn, and sloe.

There is hardly any article of food in general use

which has not somewhere been converted into a stimulant by the process of fermentation. What else are whisky, rum, beer, etc., but fermented or distilled bread, the bread-corn diverted from its legitimate use to produce an artificial stimulant? Potatoes, sugar, honey, as well as grapes, plums, apples, cherries, and innumerable other fruits, have thus been turned from a blessing into a curse. The Moors of Barbary and Tripoli distill an ardent spirit from the fruit of the date-palm, the Brazilians from the marrow of the sago-tree and from pineapples, and even the poor berries that manage to ripen on the banks of the Yukon have to furnish a poison for the inhabitants of Alaska. Pulque, the national drink of Mexico, is derived from a large variety of the aloe-plant, the sap of which is collected and fermented in buckskin sloughs into a turbid yellowish liquor of most vicious taste.

Cheese, in fact, is nothing but coagulated milk in a more or less advanced state of decay. Sauerkraut is cabbage in the first stage of fermentation, which if completed yields quass, the above-mentioned Russian tonic. Chica, a whitish liquid which in Peru is handed around like coffee, after meals, is prepared from maize or Indian corn, moistened and fermented by mastication. How a fondness for such abominations is propagated can be explained by any boy who had to drink beer or eat strong cheese against his will, and by-and-by "rather liked it," but a question less easily answered is how such tastes ever could originate. To the first man who tasted hasheesh, alcohol, or pulque, these substances could hardly be more tempting, we should think, than coal-tar or caustic sublimate. But most articles of food and drink are older than history. All we can do is to trace their progress from nation to nation and from century to century, but their origin loses itself in the cloudland of tradition. The exegesis of diet is as problematic as that of religious dogmas.

By avoiding pungent condiments we also obviate the *principal cause of gluttony*. It is well known that the admirers of lager-beer do not drink it for the sake of its nutritive properties, but as a medium of stimulation, and I hold that nine out of ten gluttons swallow their peppered ragouts for the same purpose. Only natural appetites have natural limits. Two quarts of water will satisfy the normal thirst of a giant, two pounds of dates his hunger after a two days' fast. But the beer-drinker swills till he runs over, and the glutton stuffs himself till the oppression of his chest threatens him with suffocation. Their unnatural appetite has no limits but those of their abdominal capacity. *Poisonhunger* would be a better word than appetite. What they really want is alcohol and hot spices, and, being unable to swallow them "straight," the one takes a bucketful of swill, the other a potful of grease into the bargain.

But gluttony has one other cause—involuntary cramming. Fond mothers often surfeit their babies till they sputter and spew, and it is not less wrong to force a child to eat any particular kind of food against his grain—in disregard of a natural antipathy. Such aversions are allied to the feeling of repletion by which Nature warns the eater to desist, and, if this warning is persistently disregarded, the monitory instinct finally suspends its function; overeating becomes a morbid habit, our system has adapted itself to the abnormal condition, and every deviation from the new routine produces the same feeling of distress which shackles the rum-drinker to his unnatural practice. Avoid pungent spices, do not cram your children against their will, and never fear that natural aliments will tempt them to excess. But I should add here that of absolutely innocuous food—ripe **food and** simple farinaceous preparations—a larger quantity than is commonly imagined can be habitually taken with perfect freedom from injurious consequences. On the upper Rhine they have *Trauben-Curen*—sanitaria where people are fed almost exclusively on ripe grapes in order to purify their blood. The grapes generally used for this purpose are of the variety known as Muskateller, with big, honey-sweet berries of a most enticing flavor. "Doesn't such physic tempt your patients?" I asked the manager of a famous Trauben-Curen; "don't they dose themselves to a damaging extent?" His answer surprised me. "Damaging? Yes, sir," said he, "they damage my pocket, some of them do, though I charge them three florins a day, lodgers five. They can not damage *themselves* by eating Muskateller."

Never stint the supply of **fresh** drinking-water. The danger of water-drinking in **warm weather** has been grossly exaggerated. Cold water and cold air are the two scape-goats that have to bear the burden of our besetting sins. There is, indeed, something preposterous in the idea that Nature would punish us for indulging a natural appetite to its full extent. Sheep that have been fed on dry corn-husks all winter sometimes break into a clover-field and eat till they **burst**; but who ever heard of a dyspeptic bear, or of an elk prostrated by a fit of gastric spasms? And yet we need not doubt that wild animals eat while their appetite lasts. If we lock **them up** and deprive them of their wonted exercise, their appetite, too, diminishes. In short, as long as **we confine ourselves** to our proper diet, our stomachs never call for more than we can digest. There are things that have to be eaten in homœopathic doses to prevent surfeit, but respecting such stuff (Limburger, caviare, etc., I would say, as of spices and alcohol), abstinence is better than temperance. In convivial neighborhoods sporadic cases of surfeit are almost as unavoidable as Christmas dinners and school picnics; but their efforts are as transient as their causes. For children, a nearly infallible peptic corrective is a *fast-day passed in cheerful out-door exercise*. By a curious law of periodicity, the mind will stray to the dining-room when the wonted meal-time comes around, even if genuine appetite does not return with that hour, but fishing, hunting, and ball-playing divert our thoughts from such channels, and, returning late in the evening from a good day's sport, the periodicity of bedroom-thoughts, aided by fatigue, overcomes the latent craving for food without the least effort. Try the experiment.

Want of appetite is not always a morbid symptom, nor even a sign of imperfect digestion. Nature may have found it necessary to muster all the energies of our system for some special purpose, momentarily of paramount importance. Organic changes and repairs, teething, pleuritic epurations, and the external elimination of bad humors (boils, etc.), are often attended with a temporary suspension of the alimentary process. The instinct of domestic animals thus generally counteracts the influence of abnormal circumstances. As a rule, it is always the safest plan to give Nature her own way, and was thus proved even in the extreme cases of more than one *bona fide* fasting girl, whose system, for recondite reasons of its own, preferred to subsist on air for weeks and months together.

In regard to the quality of food, too, there are intuitive dislikes which should not be disregarded, because they can not always be accounted for. I do not say *likes and* dislikes; a child's whimsical desire to treat innutritious or injurious substances as comestibles should certainly not be encouraged as long as its hunger can be appeased with less suspicious aliments. For it is a curious fact that *all* unnatural practices—the eating of indigestible matter as well as of poisons—are apt to excite a morbid appetency akin to the stimulant habit. The human stomach can be accustomed to the most preposterous things. The Otomacs, of South America, whose forefathers in times of scarcity may have filled their bellies with loam, are now afflicted with a national *penchant* for swallowing inorganic substances. In New Caledonia, *habitués* often eat as much as two pounds of ferruginous clay a day, and a similar stuff is sold in the markets of Bolivia, and finds eager purchasers, even when better comestibles are cheaper. Professor Ehrenberg procured a sample of this clay which was supposed to contain organic admixtures or some kind of fat; but his analysis proved that it consists of talc, mica, and a little oxide of iron. According to Malte-Brun, the Lisbon lazzaroni chew all day long the insipid, leathery kernels of the carob-bean (*Mimosa silica*), and the most popular "chewing-gum" is said to be composed chiefly (not entirely, I hope) of resin, paraffine, and triturated caoutchouc! Still, Ehrenberg's analysis makes stranger things credible. I do not doubt that a man might contract a habit of swallowing a couple of slate-pencils or a dime's worth of shoe-strings every morning.

But an innate *repugnance* to a special dish, or even to a special class of aliments, may be indulged very cheaply, and certainly very safely, as long as there are other available substances of the same nutritive value.

Abnormal antipathies may indicate constitutional abnormities, and among the curious cases on record there are some which clearly preclude the idea of imaginative influences. I knew a Belgian soldier on whom common salt, in any combination, and in any dose exceeding ten pennyweights, acted as a drastic poison, and thousands of Hindoos can not taste animal food without vomiting. Similar effects have obliged individuals to abstain from onions, sage, parsnips, and even from Irish potatoes. Dr. Pereira mentions the case of an English boy who had an incurable aversion to mutton: "He could not eat mutton in any form. The peculiarity was supposed to be owing to caprice, but the mutton was repeatedly disguised and given to him unknown; but uniformly with the same result of producing violent vomiting and diarrhœa. And from the severity of the effects, which were in fact those of a virulent poison, there can be little doubt that, if the use of mutton had been persisted in, it would soon have destroyed the life of the individual."*

It may be considered as a suggestive circumstance that the great plurality of such instinctive aversions relate either to stimulants or to some kind of animal food. To one person whose stomach can not bear bread or apples, we shall find a thousand with an invincible repugnance to pork, coffee, and pungent condiments. It is also certain that, by voluntary abstinence from all such things, the vigor of the alimentary organs can be considerably increased. The Danish sailors whom the Dey of Algiers had fed on barley and dates for a couple of months, found that after that they "could digest almost anything." †

* Pereira, "Treatise on Food and Diet," p. 242.
† Wodderstadt, "On Yellow Fever," p. 72.

By adopting an absolutely non-stimulating, chiefly vegetable diet, combined with active exercise in open air, the most dyspeptic glutton can cure himself in the course of a single season, and by the same means every boarding-school might become a dietetic sanitarium. The following list of hygienic *menus* is arranged in the order of their digestibility and wholesomeness:

Milk, bread, and fruit.—Eggs (raw or whipped), bread and honey.—Boiled eggs, bread, and apples (ancient Rome).—Bread and butter, rice-pudding, with sugar and fresh milk.—Corn-bread or roasted chestnuts, butter, honey, and grapes (the usual diet of the long-lived Corsican mountaineers).—Fish, butter, oatmeal-porridge, and fresh milk (Danish Islands).—Pancakes, honey or new molasses, poached eggs, boiled milk, and bread-pudding.—Vegetable soups, baked beans, potatoes (baked or mashed), butter, biscuits, and apple-dumplings.

GENERAL RULES.—Avoid stimulants; alcoholic and narcotic drinks, tobacco, and all pungent spices; be sparing in the use of animal food, especially in summertime; in midsummer eat fruit with every meal; let unprepared food (fresh milk, fruits, etc.) form a part of your daily fare; of unprepared aliments, as well as of all unspiced viands, the most palatable are the most wholesome; eat slowly and masticate your food; never eat if you have no appetite; and finish your last meal three hours before bed-time.

As a dessert I will add a few of my favorite dietetic aphorisms: An hour of exercise to every pound of food.—We are not nourished by what we eat, but by what we digest.—Every hour you steal from digestion will be reclaimed by indigestion.—Beware of the wrath of a patient stomach!—He who controls his appetite in regard to the quality of his food may safely indulge it in regard to quantity.—The oftener you eat, the oftener you will repent it.—Dyspepsia is a poor pedestrian; walk at the rate of four miles an hour, and you will soon leave her behind.—The road to the rum-cellar leads through the coffee-house.—Abstinence from *all* stimulants, only, is easier than temperance.—There are worthier objects of charity than famine-stricken nations that send their breadstuffs to the distillery.—An egg is worth a pound of meat; a milch-cow, seven stall-fed oxen.—Sleep is sweeter after a fast-day than after a feast-day.—For every meal you lose you gain a better.

How often should we eat is still a mooted question. For men in a state of nature the answer would be simple enough; but, considering our present artificial modes of life, I must say that the choice of fixed hours is less important than the observation of the following rule: *Never eat till you have leisure to digest.* For digestion requires leisure; we can not assimilate our food while the functional energy of our system is engrossed by other occupations. After a hearty feed, animals retire to a quiet hiding-place; and the "after-dinner laziness," the plea of our system for rest, should admonish us to imitate their example. The idea that exercise after dinner promotes digestion is a mischievous fallacy; Jules Virey settled that question by a cruel but conclusive experiment. He selected two curs of the same size, age, and general *physique*, made them keep a fast-day and treated them the next morning to a square meal of potato-chips and cubes of fat mutton, but, as soon as one of them had 'eaten his fill, he made the other stop too, to make sure that they had both consumed the same quantity. Dog No. 1 was then confined in a comfortable kennel, while No. 2 had to run after the doctor's coach, not at a breathless rate of speed, but at a fair, brisk trot, for two hours and a half. As soon as they got home, the coach-dog and his comrade were slain and dissected; the kennel-dog had completely digested his meal, while the chips and cubes in the coach-dog's stomach had not changed their form at all; the process of assimilation had not even begun! Railroad laborers, who bolt their dinner during a short interval of hard work, might as well pass their recess in a hammock; instead of strengthening them, their dinner will only oppress them, till it is digested, together with their supper, in the cool of the evening. In a manner essentially similar, mental activity tends to hinder the digestive process for a considerable time; and I believe, more especially, the digestion of the very substances that are often selected as brain-food *par excellence*. Even after a fashionable dinner of six or seven courses (*curses*, Dr. Abernethy used to call them), two hours of absolute rest will set our wits a-work again; but, if that time be passed behind a double-entry ledger, a feeling of lassitude, often combined with an almost resistless somnolence, will advise the brain-worker that his vital energy is needed for other purposes. "I could eat with more comfort if it wasn't for the consciousness of having to hurry back to my drudgery," I heard a poor class-teacher say, and the same consciousness embitters the noonday-meal of millions of school-children and overworked clerks.

Andrew Combe, M. D., informs us that a century ago the tradesmen of Edinburgh used to indulge in a "nooning," a general suspension of business for two hours, in the middle of the day. But an hour or so was thus probably spent in going home and back, dressing, etc., and half an hour at the meal itself; so that, after all, only thirty minutes remained for digestion; and, considering the anachronism of that nooning practice, the best plan, on the whole, would seem to be a general return to the method of the ancient Romans, who postponed their principal meal until their day's work was done. It would be an insult to common sense and humanity to doubt that the eight-hour system will ultimately prevail, and, where it has been already adopted, I can see no reason why mechanics could not arrange to finish their day's job at 4 P. M. Schools should always close at four. Bankers and government clerks often get home before that time, and competitive shopkeepers might carry on their business by relays. At half-past four, or, say, five o'clock, the *coena domestica* might begin, conclude before six; then *dolce far niente*, pleasant conversation, and four blessed hours for digestion.

But that principal meal should be the last. It is an important rule that we should digest our food thoroughly before we replenish the stomach. To counteract the effects of over-eating, the gluttons of ancient Rome used emetics, the Parisian gastronomes stimulants. Dr. Alcott wants us to "leave off hungry"; the exponents of the movement-cure prescribe a certain system of gymnastic evolutions before and after dinner. But there is a better plan: *Lengthen the interval between meals.* Two meals a day are enough, perhaps more than enough, though we can accustom ourselves to swallow (not digest) five or six. It all depends on training, and in no other respect is the human system so plastic to the influence of habit. The Rev. Mr. Moffat tells us that the Gonaque Hottentots are noways incommoded by a five days' fast, and get old on an average of four meals a week. The Greeks and Romans during the prime of their republic contented themselves with one meal a day; Claude Bernard recommends two, but his countrymen generally eat three; their German neighbors four; the East-Germans even five: breakfast, second breakfast (*zweites Frühstück*), dinner, *Vesperbrot*, and supper, to which supper the Vienna burghers actually superadd a *Nacht-bissel*—a "night-lunch," of cold potato-salad with bread and *Wurst*, and often with a mug of beer—"for the stomach's sake"! I get along comfortably with a meal and a half; so does my granduncle, an octogenarian, who still masticates his bread with a full set of unbought teeth. Two, or one and two halves, should be enough for any man. The lightest breakfast is the best—buckwheat-cakes with a little honey or apple-butter, and a glass of milk, or a cup of chocolate, if you must take "something warm." Chocolate possesses nutritive properties, which tea and coffee *per se* are totally devoid of. I never use it, but I believe it is non-stimulating. Or chew a crust of stale bread, the best dentifrice and a useful absorbent, good for acidity of the stomach. At noon take a glass of milk and a couple of biscuits, or in summer a couple of ripe pears or peaches; they will keep you cool during the post-meridian heat, and do you more good than a cocktail lunch. Never keep a pocket-flask. Don't stay with flagons; better comfort with apples if you can not wait till five. School-children should pass their recess on the play-ground. A biscuit and a pocketful of apples will satisfy the temporary demands of the stomach; and, if they have munched up their comestibles in the course of the morning, as boys are apt to do, they will find it far easier to forego their noonday lunch altogether than to resist the insidious somnolence which would dull their wits after a regular dinner, and often makes the afternoon lesson a protracted struggle between nature and duty.

But at the principal meal they should eat their fill. Let them pitch in, without fear of dangerous consequences—unless your landlord charges by the plateful. Children, like monkeys, have a way of dallying with their food if they are full—picking a crumb here and there, or mumbling their apples without using their teeth. Make them get up if you notice such symptoms, or, better, entice them away by improvising some outdoor or up-stairs amusement. But I repeat, never press them to eat—for principle's sake—not even your young visitors; they are not likely to go to bed hungry if your *menu* comprises such items as baked apples or bread-pudding and sweet milk.

Jean Jacques Rousseau holds that intemperate habits are mostly acquired in early boyhood, when blind deference to social precedents is apt to overcome our natural antipathies, and that those who have passed that period in safety have generally escaped the danger of temptation. The same holds good of other dietetic abuses. If a child's natural aversion to vice has never been willfully perverted, the time will come when his welfare may be intrusted to the safe-keeping of his protective instincts. You need not fear that he will swerve from the path of health when his simple habits, sanctioned by Nature and inclination, have acquired the additional strength of long practice. When the age of blind deference is passed, vice is generally too unattractive to be very dangerous. "Why make yourself the slave of such a degrading habit?" says Count Zinzendorf, in his "Hirtenbrief"; "it is so easy never to begin!" I go further. I say it is difficult to begin. Nature is not neutral on a point of such importance.

Between virtue and vice she has erected a bulwark which she intended to last from birth to death. We need not strengthen that bulwark. We need not guard it with anxious care; it will stand the ordinary wear and tear of life. All we have to do is to save ourselves the extraordinary trouble of breaking it down.

Pure joys never pall; uniformity is uniform happiness if the even tenor of our way is the way of Nature. And Nature herself will guide our steps if the exigence of abnormal circumstances should require a deviation from the beaten path. Remedial instincts are not confined to the lower animals; man has his full share of them; the self-regulating power of the human system is as wonderful in the variety as in the simplicity of its resources. Have you ever observed the weather-wisdom of the black bind-weed?—how its flowers open to the morning sun and close at the approach of the noontide glare; how its tendrils expand their spirals in a calm, but contract and cling, as with hands, to their support when the storm-wind sweeps the woods? With the same certainty our dietetic instincts respond to the varying demands of our daily life. Without the aid of art, without the assistance of our own experience, they even adapt themselves to the exigencies of our abnormal social conditions, and our interference alone often prevents them from counteracting the tendency of dire abuses.

Summer brings no repose to the slaves of Mammon, but dull headaches and the stomach's imperative demand for rest convince even the unwilling that intricate arithmetical problems and 90° Fahr. are incompatible with digestion; and I ascribe it to the logic of those gastric arguments that bankers and brokers now close their shops at 3 P. M.; and that business-men generally avoid repletion in the middle of the day. "Cheese is gold in the morning, silver at noon, and lead at night," says a mediæval proverb; but the effects of those horrid cheese and porter breakfasts of Queen Anne's time satisfied our grandams that rotten curd and fermented (i. e., putrid) barley-broth are always lead, except to those who employ the hygienic philosopher's stone—active and long-continued out-door exercise. After recovery from an exhausting sickness—especially if you decide to promote that recovery by throwing physic to the dogs—the demands of your stomach will often become exorbitant, but only apparently so; your system wants to repair the waste of the disease. Never fear that "the digestive organs are too feeble yet," etc.; those organs will keep their promise, unless you break yours by resuming medication. Have you eaten more than the wants of your system require? Your appetite will not respond to your invitation at the next meal. Take the hint—wait. Do not increase the troubles of your stomach by mordant spices and alcohol. In the sultry dog-days your system craves a surcease of greasy ragouts and yearns for something refreshing—sherbet or cool fruit. Get a water-melon. "But isn't the yellow fever in town? Quack, Quinine, and other leading physicians, agree that one must take a course of antiseptics, and avoid vegetables at such seasons." Don't believe them; Nature knows better. Fruit is a better antiseptic than fusel poison and wormwood. The frugivorous Mexican survives where the beef-eating stranger dies in spite of his bitters. If sailors have been surfeited with salt meat, their craving after lemon-juice or fresh fruit becomes more urgent from day to day; the surcharge of their organism with saline matter requires a neutralizing acid. A single meal of salt herring excites merely thirst; common water is yet sufficient to dilute the ingesta and eliminate the salt. Vegetable substances that consist chiefly of starch and water supply the wants of our organism less completely than those that contain an admixture of gluten, albumen, and fat; and, if we restrict our diet to the first-named class of aliments, our system announces the deficit by means of our senses; without such complements as milk, sugar, or fat, rice-bread is more insipid than bread from unbolted wheat-flour.

All dietetic needs of our body thus announce themselves in a versatile language of their own, and he who has learned to interpret that language, nor willfully disregards its just appeals, may avoid all digestive disorders—not by fasting if he is hungry, or forcing food upon his protesting stomach, not by convulsing his bowels with nauseous drugs, but by quietly following the guidance of his instincts.

Nature's health laws are simple. The road to health and happiness is not the labyrinthine maze described by our medical mystagogues. In perusing their dietetic codes one is fairly bewildered by a mass of incongruous precepts and prescriptions, laborious compromises between old and new theories, arbitrary rules, and illogical exceptions, anti-natural restrictions and anti-natural remedies. Their view of the constitution of man suggests the King of Aragon's remark about the cycles and epicycles of the Ptolemaic system: "It strikes me the Creator might have arranged this business in a simpler way."

All normal things are good, all evil is abnormal, is an axiom which has been almost reversed in the principle of our orthodox health theories, for many of our physical educators still hold to the cardinal error of their spiritual colleagues, who consider depravity and wretchedness as the normal condition of man, and happiness as the reward of a self-abhorring suppression of all natural desires and of a blind confidence in the efficacy of an abnormal and mysterious remedy—nay, who despise Earth herself as a "vale of tears," and life as a disease whose only cure is death, whose only anodyne a dream of a supernatural elysium. It is time to awake from that dream. It is time to open our eyes to the well-springs of life and happiness which the bounty of our Mother Earth sends forth in such abundance, and which man might enjoy with all his fellow-creatures if his perversity had not turned them into sources of misery and death. Instead of insulting our Maker by the doctrine of innate depravity, we should learn to distinguish the voice of our natural instincts from the cravings of a morbid appetency. We should try to restore life to its original purity and healthfulness, instead of despising it and looking for happiness beyond the grave.

But the deluge of mediæval superstitions is fast assuaging, and many a submerged truth has reappeared like a bequest of a former and better world, and now stands as a way-mark on the road to a true Science of Life. We have rediscovered the truth that the weal and woe of earth are not distributed by the caprices of a mysterious Fate, but that they follow as sure effects upon ascertainable causes. Our best thinkers have ceased to doubt that man can work out his own destiny, that the Creator has made us the keepers of our own happiness on conditions which he never violates; that he has attached pleasure to every right act, and pain to every wrong, that he fulfills the promises of our yearnings, and never permits us to sin unwarned. We have at last begun to realize the fact that the physical laws of God find an echo in the voice of our innate monitor, and only an hereditary mistrust in our instincts makes us still hesitate to commit ourselves to its guidance. But experience will overcome that prejudice by-and-by; duty and inclination will go hand in hand, and the result will justify our trust in the wisdom and benevolence of Nature.

CHAPTER II.

IN-DOOR LIFE.

"What is to the mind a healthy body,
To the body is a healthy house."
—Fabio Colonna.

NEXT to our dietetic sins, the abuses connected with our habits of domestic life have contributed the largest share to the great sum of human misery. Yet few evils might be more easily avoided. There are diseases which may be considered as visitations of national iniquities whose consequences are almost beyond the control of individuals; but for the sufferings caused by scrofula and pulmonary disorders we are indebted chiefly to our own prejudices. Prejudice and ignorance have filled more consumptives' graves than poverty. Even in large manufacturing towns air is free. If our artisans could realize the consequences of breathing miasma, they would prefer the life-air of the wildest wilderness to the lung-poison of their slums; like a caged bird, the tenement prisoner would refuse to pair rather than people the earth with cachectic wretches. The exodus of their workmen would soon induce manufacturers to imitate the founder of Saltaire; building speculators would find it to their adventage to adopt the Philadelphia plan, adding suburb to suburb rather than loft to loft; cities would grow outward instead of upward. A reform of that sort would imply various modifications of our present labor system; but before the enlightenment of public opinion such difficulties vanish like mist before the rising sun. There was a time when it was actually proposed to abolish the summer vacations of the French town schools "in order to enlarge their curriculum in proportion to the advance of modern science"; but, since we have ascertained that out-door exercise is more important than all the graphies and ologies of the Académie Française, it has been found that, with a well-arranged plan of instruction ten months a year, five days a week and six hours a day are quite enough for any school. If the eight-hour system were generally adopted, operatives would not be compelled to live within ear-shot of the factory-whistle, and in very large cities the daily influx and reflux of a suburban multitude would enable railroad companies to carry individuals at rates which the poorest would call moderate. Far enough from the city center to evade the region of dear building-lots, and yet within easy reach of all kinds of door and sash factories and planing-mills, there would be no need of crowding three generations into a single room, and suffocating them with mingled kitchen-fumes and sick-bed odors. Three rooms and an out-house should be the minimum for a family with children.

In a tolerable location, the air of a three-room cottage can be kept pure enough without force ventilators or any other expensive contrivance. Open your windows; in very cold weather, air the bedrooms in day-time and the others at night. In larger houses, the kitchen, parlor, and dining-room should be thoroughly ventilated every night, also in day-time at convenient intervals, during the temporary absence of the occupants. To save foul air for the sake of its warmth is poor economy; experiments would show that the difference in fuel amounts only to a trifle, anyhow. Ten or twelve pounds of coal a day ought not to weigh against the direct gain in comfort and the prospective, unspeakable gain in health. Breathing the same air over and over again means to feed the organism on the excretions of our own lungs, air surcharged with noxious gases and almost depleted of the life-sustaining principle. Azotized air affects the lungs as the substitution of excrements for nourishing food would affect our digestive organs: corruption sets in; pulmonary phthisis is, in fact, a process of putrefaction.

No ventilatory contrivance can compare with the simple plan of opening a window; in wet nights a "rain-shutter" (a blind with large, overlapping bars) will keep a room both airy and dry. In every bedroom, one of the upper windows should be kept open night and day, except in storms, accompanied with rain or with a degree of cold exceeding 10° Fahr. In warm summer nights open every window in the house and every door connecting the bedroom with the adjoining apartments. Create a thorough draught. Before we can hope to fight consumption with any chance of success, we have to get rid of the *night-air superstition*. Like the dread of cold water, raw fruit, etc., it is founded on that mistrust of our instincts which we owe to our anti-natural religion. It is probably the most prolific single cause of impaired health, even among the civilized nations of our enlightened age, though its absurdity rivals the grossest delusions of the witchcraft era. The subjection of holy reason to hearsays could hardly go further.

"Beware of the night-wind; be sure and close your windows after dark"! In other words, beware of God's free air; be sure and infect your lungs with the stagnant, azotized, and offensive atmosphere of your bedroom. In other words, beware of the rock spring; stick to sewerage. **Is night-air injurious?** Is there a single tenable pretext for such an idea? Since the **day of creation that air has been breathed with impunity by millions of different animals**—tender, **delicate creatures, some of them—fawns, lambs, and young birds. The moist night-air of the tropical forests is breathed** with impunity by our next relatives, the anthropoid apes— the same apes that soon perish with consumption in the close though generally **well-warmed** atmosphere of our northern menageries. Thousands of soldiers, hunters, **and lumbermen sleep** every night **in tents** and open **sheds without the least** injurious consequences; men **in the last stage of consumption have recovered by adopting a** semi-savage **mode of** life, **and** camping out-doors **in all but the stormiest nights. Is it** the draught you **fear, or the** contrast **of temperature?** Blacksmiths and railroad-conductors seem to thrive under such influences. Draught? **Have you never seen boys skating in the teeth of a snow-storm at the rate of fifteen miles an hour? "They counteract the effect of the cold air by vigorous exercise." Is** there no other **way of keeping warm? Does the north wind** damage the fine lady sitting motionless in her sleigh, or the pilot and helmsman of a storm-tossed **vessel?** It can not be the *inclemency* of the open air, for, even in sweltering summer nights, the sweet south wind, blessed by **all** creatures **that draw the breath of life, brings no relief to the victim** of aërophobia. There is **no doubt that families who** have **freed themselves from the curse of** that superstition can **live out and out healthier in the heart of a great city than its** slaves **on the airiest** highland of the southern **Apennines.**

People of sedentary habits can actually become fond of foul air. In our large cities thousands of in-door laborers are afflicted with what I might call the *troglodyte habit*. The troglodytes, or cave-dwellers of ancient Nubia, belonged to a tribe which seems to have formed an intermediate link between the Semitic and Ethiopian races, but which had become entirely extinct before the second century of the Christian era. Between Sidi Elgor and Port Er-nassid (the ancient Berenice), on the shores of the Red Sea, a German traveler examined many of the limestone-caverns which were the favorite haunts of these singular beings, and found no difficulty in distinguishing the bones of the Coptic and Arabian burial-places from the troglodyte skeletons, which could be recognized by their demi-simian skulls, their attenuated brachial and femoral bones, and especially their narrow chests.

These peculiarities he ascribes to the unnatural habits of the wretched cave-men, who, from cowardice or constitutional sloth, passed the greater part of their existence in the penetralia of their foul burrows, while their neighbors preferred a manlier way of securing themselves against enemies and wild beasts, and saved themselves from the glow of the midsummer sun by cultivating shade-trees. "Herodotus speaks of persecutions," he remarks, "but this fixed custom of theirs may, perhaps, be attributed to vicious habit, strengthened by hereditary transmission, quite as much as to necessity, for men can become fond of vitiated air, as they contract a passion for fermented drink or decayed food." In 1853, when Hanover and other parts of Northern Germany were visited by a very malignant kind of small-pox, the great anatomist, Langenbeck, tried to discover "the peculiarity of organic structure which disposes one man to catch the disease while his neighbor escapes. . . . I have cut up more human bodies than the Old Man of the Mountain with all his accomplices," he writes from Göttingen in his semi-annual report, "and, speaking only of my primary object, I must confess that I am no wiser than before. But, though the mystery of small-pox has eluded my search, my labors have not been in vain; they have revealed to me something else—the origin of consumption. I am sure now of what I suspected long ago, viz., that pulmonary diseases have very little to do with intemperance or with erotic excesses, and much less with cold weather, but are nearly exclusively (if we except tuberculous tendencies inherited from *both* parents, I say *quite* exclusively) produced by the breathing of foul air. The lungs of all persons, minors included, who had worked for some years in close workshops and dusty factories, showed the germs of the fatal disease, while confirmed inebriates, who had passed their days in open air, had preserved their respiratory organs intact, whatever inroads their excesses had made on the rest of their system. If I should go into practice and undertake the cure of a consumptive, I should begin by driving him out into the *Deister* (a densely-wooded mountain-range of Hanover), and prevent him from entering a house for a year or two."

The ablest pathologists of the present time incline to the same view. "There *is* a cure for consumption," says Dio Lewis, "though I doubt if it will ever become popular. Even in its advanced stages the disease may be arrested by *roughing it;* I mean by adopting savage habits, and living out-doors altogether, and in all kinds of weather."

That low temperature in open air does not injure

our lungs has been recognized even by old-school physicians, who now send their patients to Minnesota and Northern Michigan quite as often as to Florida; and is conclusively proved by the fact that of all nations of the earth, next to the inhabitants of the Senegal highlands, the Norwegians, Icelanders, and Yakuts of Northern Siberia, enjoy the most perfect immunity from tubercular diseases. Dry and intensely cold air preserves decaying organic tissue by arresting decomposition, and it would be difficult to explain how the most effective remedy came to be suspected of being the cause of tuberculosis, unless we remember that, where fuel is accessible, the disciple of civilization rarely fails to take refuge from excessive cold in its opposite extreme—an overheated artificial atmosphere—and thus comes to connect severe winters with the idea of pectoral complaints.

There is a rather numerous class of beasts whose lungs seem able to adapt themselves to an atmosphere almost devoid of oxygen, but the human animal and the Quadrumana do not belong to that class. Monsieur de la Motte-Baudin, who was connected with the scientific staff of the *Jardin des Plantes* as their "menagerie-doctor" for more than twenty years, never omitted to dissect his deceased patients before turning them over to the taxidermist, and invariably found that *all* monkeys had succumbed to some variety of phthisis, while the lungs of the badgers, bears, and foxes, were perfectly sound. The three last-named animals are natural cave-dwellers, and have been provided with organs especially contrived to resist the effluvia of their burrows; while the *Simiæ*, like man, are open-air creatures, whose proper atmosphere is the cordial air of woodlands.

Among the natives of Senegambia pulmonary affections are not only nearly but absolutely unknown; yet a single year passed in the overcrowded man-pens and steerage-hells of the slave-trader often sufficed to develop the disease in that most virulent form known as galloping consumption; and the brutal planters of the Spanish Antilles made a rule of never buying an imported negro before they had "tested his wind," i. e., trotted him up-hill and watched his respirations. If he proved to be "a roarer," as turfmen term it, they knew that the dungeon had done its work, and discounted his value accordingly. "If a perfectly sound man is imprisoned for life," says Baron d'Arblay, the Belgian philanthropist, "his lungs, as a rule, will first show symptoms of disease, and shorten his misery by a hectic decline, unless he should commit suicide."

Our home statistics show that the percentage of deaths by consumption in each State bears an exact proportion to the greater or smaller number of inhabitants who follow in-door occupations, and is highest in the factory districts of New England and the crowded cities of our central States. In Great Britain the rate increases with the latitude, and attains its maximum height in Glasgow, where, as Sir Charles Brodie remarks, windows are opened only one day for every two in Birmingham, and every three and a half in London; but going farther north the percentage suddenly sinks from twenty-three to eleven, and even to six, if we cross the fifty-seventh parallel, which marks the boundary between the manufacturing counties of Central Scotland and the pastoral regions of the north.

It is distressingly probable, then, to say the least, that consumption, that most fearful scourge of the human race, is *not* a "mysterious dispensation of Providence," nor a "product of our outrageous climate," but the direct consequence of an outrageous violation of the physical laws of God. Dyspepsia (for which also open-air exercise is the only remedy), hypochondria, and not only obstruction but destruction of the sense of smell—"knowledge from one entrance quite shut out"—will all be pronounced mere trifles by any one who has witnessed the protracted agony of the *Luft-Noth*, as the Germans call it with horrid directness—the frantic, ineffectual struggle for life-air. Dr. Haller thought that, if God punishes suicide, he would make an exception in favor of consumptives; and there is no doubt that, without the merit of martyrdom, the victim of the cruel disease endures worse than ever Eastern despot or grand-Inquisitor could inflict on the objects of his wrath, because the same amount of torture in any other form would induce speedier death.

But not only the punishments but also the warnings of Nature are proportioned to the magnitude of each offense against her laws. Injurious substances are repulsive to our taste, incipient exhaustion warns us by a feeling of hunger or weariness, and every strain on our frame that threatens us with rupture or dislocation announces the danger by an unmistakable appeal to our sensorium. How, then, can it be reconciled with the immutable laws of life that the greatest bane of our physical organism overcomes us so unawares that consumption is proverbially referred to as the insidious disease? Should it really be possible that Nature has failed to provide any alarm-signals against a danger like this? The truth is, that none of her protests are more pathetic or more persistent than those directed against the habit that is fraught with such pernicious consequences to our respiratory organs.

It is probable that some of the victims of our numerous dietetic abuses have become initiated to these vices at such an early period of their lives that they have forgotten the time when the taste of tea and alcohol seemed bitter, or the smell of tobacco produced nausea; but I am certain that no man gifted with a moderate share of memory, who has grown up in the pest atmosphere of our city tenements, school-rooms, and workshops, can forget the passionate yearnings of his childhood for the free air of the woods and mountains; the wild outcry of his instinct against the process that inoculated him with the seeds of death, and stunted the development of his most vital faculties. The remorselessness of the pagan Chinese, who smother the life-spark of their infants in the swift embrace of the river-god, is mercy itself compared to the cruelty of Christian parents who suffocate their children by the slow process of stinting their life-air, through years and years of confinement in dungeons to which an enlightened community would not even consign their malefactors.

Honest Jean Paul relates that he used to secure a seat in a certain corner of an overcrowded village school-house, where a knot-hole in the wall established a communication with the outer world. Through this orifice he imbibed comfort and inspiration as from a flask, but conceived conscientious scruples against the practice, as he never could indulge without becoming conscious of a temptation to abandon his old parents and his home, and join a troop of wood-cutters or gypsies, not from any vagrant tendencies, or want of dutiful sentiments, but from an almost irresistible desire to make the luxury of fresh air a permanent blessing. "I knew they would charge me with black ingratitude, if I should run away," he says. "Good God! how I longed to prove my affection by working for them in wind and weather, fetching in cord-wood from the woods and splitting it into the nicest, handiest pieces, carrying messages over the snow-covered mountains and be back in half the time any one else could make the trip—do anything that would save me—not from my books, but from that glowing Moloch of a big stove, and that stifling, soul-stifling smell of our dungeon!"

Even to the most inveterate believer in natural depravity this might suggest a doubt whether the repugnance of children to study may not be founded on a physical virtue rather than on moral perverseness. To whatever is really beneficent we are commonly drawn by natural attraction, and whatever appears violently repulsive to youthful minds may be justly suspected of containing more of evil than of good. The very disciple of Socrates who used to run sixteen miles a day to hear the ἄριστος ἰατρῶν (best of physicians), would have hesitated to purchase physic for his soul at the price of physical health; and we can not blame our children for being unable to reconcile the precepts they hear with those they feel, and giving way now and then to the more consistent and more logical prompter.

The farmer's boy may look forward to each afternoon and each summer vacation as a refreshing interlude, and to the last term of his school-years as the last act of the tragedy; but in cities the end of the schoolroom bondage is too often the beginning of the endless slavery which awaits the young apprentice of the workshops, factories, and counting-houses. In Northwestern Europe and the Eastern States of North America, eleven million human beings, a fourth of that number minors, are performing their daily toil in an atmosphere that saps the vigor of their souls and bodies more effectually than a diet of potatoes and water could do it in the same time. A full third of the cotton-spinners of Lancashire and Massachusetts are girls and boys in their teens! They do not complain to a stranger, unless he should be able to interpret the language of their haggard faces and weary eyes; but no one who has fathomed the depth of their misery will charge me with exaggeration if I say that, to the vast majority of the unfortunates, loss of feeling and of reason would be a blessing. What *do* they feel but unsatisfied hunger in a hundred forms, and what can reason tell them but that they have been defrauded of their birthright to happiness; that not only their opportunity but their capacity for enjoyment is ebbing away; and that, whatever after-years may bring, their life has been robbed as a day of its morning or a year of its spring-time?

The opium-habit may be acquired in less than half a year, and the natural repugnance to alcohol and tobacco is generally overcome after four or five trials; but the factory-slave has to pass through ten or fifteen years of continual struggle against his physical conscience, before the voice of instinct at last becomes silent, and the painful longing for out-door life gives way to that anæsthesia by which Nature palliates evils for which she has no remedy. In more advanced years the habit becomes confirmed, and we find old *habitués* who actually enjoy the effluvia of their prisons, and dread cold air and "draughts" as they would a messenger of death. They avoid cold instead of impurity, just as tipplers on a warm day imagine that they would "catch their death" by a draught from a cool fountain, but never hesitate to swallow the monstrous mixtures of the liquor-vender.

Rousseau expresses a belief that any man, who has preserved his native temperance for the first twenty-five years, will afterward be pretty nearly proof against temptation, because very unnatural habits can only be acquired while our tastes have the pliancy of immaturity, and I think the same holds good of the troglodyte-habit: no one who has passed twenty or twenty-five years in open air can be bribed very easily to exchange oxygen for miasma.

Shamyl-ben-Haddin, the Circassian hero chieftain, who was captured by the Russians in the winter of 1864, was carried to Novgorod and imprisoned in an apartment of the city armory, which resembled a comfortable bed-chamber rather than a dungeon, and was otherwise treated with more kindness than the Russians are wont to show their prisoners, as the Government hoped to use his influence for political purposes. But a week after his arrival in Novgorod the captive mountaineer demanded an interview with the commander of the armory, and offered to resign his liberal rations and subsist on bread and cabbage-soup like the private soldiers of his guard, and also to surrender some valuables he had concealed on his person, on condition that they would permit him to sleep in open air. One more week of such nausea and headache as the confinement in a closed room had caused him would force him to commit suicide, he said, and, if his request was refused, God would charge the guilt of the deed on his tormentors. After taking due precautions against all possibility of escape, they permitted him to sleep on the platform in front of the guard-house; and Colonel Darapski, the commander of the city, informed his government in the following spring that the health and general behavior of his prisoner were excellent, but he had slept in open air every one of the last hundred nights, with no other covering but his own worn-out mantle, and a woolen cap he had purchased from a soldier of the guard to keep his turban from getting soiled by mud and rain.

General Sam Houston, the liberator of Texas, who had exiled himself from his native State in early manhood, and passed long years not as a captive, but as a voluntary companion of the Cherokee Indians, was ever afterward unable to prolong his presence in a crowded hall or ill-ventilated room beyond ten or twelve minutes, and described his sensation on entering such a locality as one of "uneasiness, increasing to positive alarm, such as a mouse may be supposed to feel under an air-pump."

The cause of this uneasiness is less mysterious than our nature's wonderful power of adaptation that can help us ever to overcome it. The elementary changes in the human body are going on with such rapidity that the waste of tissue and organic fluids is only partially retrieved by the digestible part of the substances which we feed to the abdominal department of our laboratory twice or thrice in twenty-four hours. The difference is made up by the labors of the upper or pectoral department, which renews its supply of raw material independently, or even in spite of our will, twenty times per minute, or 70,000 times in twenty-four hours! With every breath we draw we take into our lungs about one pint of air, so that the quantity of *gaseous food* thus consumed by the body amounts in a day to 675 cubic feet. The truth, then, is that eating and drinking may be considered as secondary or supplementary functions in the complicated process performed by that living engine called the animal body, while the more important task falls to the share of the lungs. The stomach may suspend its labors entirely for twenty-four hours without serious detriment to the system, and for two or three days without endangering life, while the work of respiration can not be interrupted for six minutes without fatal consequences.

The first object of respiration is to introduce elements needed in the preparation of blood, the second to remove gaseous carbon and other secretions of the air-cells. The deleterious consequences, therefore, of breathing the same air over and over again arise not only from the exhaustion of oxygen, but also from the circumstance that the confined atmosphere may become azotized or surcharged with carbon to the limit of its absorbing powers, just as water, after being saturated with certain percents of salt or sugar, refuses to dissolve any further additions. The act of reinspiring air, which has already been subjected to the process of pulmonary digestion, is thus precisely analogous to the act of a famished animal devouring its own fæces, and if performed habitually can not fail to be attended with equally ruinous consequences. Corruption of the alimentary ducts would surely ensue in the latter (supposed) case, putrefaction of the respiratory organs *does* follow in the other. Working-men employed in localities whose azotized atmosphere is loaded besides with particles of flying cotton-fiber, metallic dust, or fatty vapors, inspire substances which are just as indigestible to their lungs as mercury and alcohol are to their stomachs, and like these cause a rapid deterioration of the tissues in proximity to which they are deposited.

The only wonder, then, is how Nature can resist outrages of this kind for any length of time; and it is a curious reflection to think what amounts of hardship of the primitive sort, such as hunger, fatigue, cold, heat, deprivation of sleep, etc., a healthy savage might accustom himself to, if he tried as hard as the poor children of civilization try to wean themselves from their hunger after life-air !

Can necessity be—we will not say an excuse, but—an explanation of such systematic self-ruin? We must utterly refuse to believe it. Wherever men barter life for bread, there is a violent presumption that they do not know what they are doing; for against recognized health-destroyers even the poorest of the poor will rebel with a promptitude that vindicates the dignity of human nature under the most abject conditions of bondage. Let a railroad contractor be caught in the trick of adulterating his flour with chalk or his sugar with alum, and see how quickly his navvies will leave him; or observe how firmly reckless Jack Tar insists on his anti-scorbutic raspberry-vinegar! Miners have left a colliery *en masse*, because the owner shirked his duty of providing safety-lamps; and the very negro slaves of a South Carolina plantation attempted the life of their master, who stinted their allowance of quinine brandy which his father had issued them to counteract the miasmatic tendencies of the rice-swamp.

Neither is it possible to suppose that want of hygienic education can be the origin of such ignorance; for Nature does not wait for the scientist to inform her children on questions of such importance. All normal things are good, all evil is abnormal; vice is a consequence of ignorance only in so far as it is a result of perverse education, and the troglodyte-habit is the direct offspring of mediæval monachism. Until after the fourth century of the Christian era, habitual indoor life between closed walls was known only as the worst form of punishment. Though the Greeks and Romans were familiar with the manufacture of glass, they never used it to obstruct their windows; in all the temples, palaces, and dwelling-houses of antiquity, the apertures provided to admit light admitted fresh air at the same time. The *tuguria* of the Roman peasants were simply arbors; and the domiciles of our hardy Saxon forefathers resembled the log-cabins of Eastern Tennessee—rough-hewed logs laid crosswise, with liberal interspaces that serve as windows on all sides except that opposed to the prevailing wind, north or northwest, where they are stopped with moss.

Men had to be utterly divorced from Nature before they could prefer the hot stench of their dungeons to the cool breezes of heaven, but our system of ethics has proved itself equal to the task. For eighteen hundred years our spiritual guides have taught us to consider Nature and everything natural as wholly evil, and to substitute therefor the supernatural and the artificial, in physical as well as in moral life. The natural sciences of antiquity they superseded by the artificial dogma, suppressed investigation to foster belief, substituted love of death for love of life, celibacy for marriage, the twilight of their gloomy vaults for the sunshine of the Chaldean mountains, and their dull religious "exercises" for the joyous games of the *palæstra*. This system taught us that the love of sport and out-door pastimes is wicked, that the flesh has to be "crucified" and the buoyant spirit crushed to make it acceptable to God; that all earthly joys are vain; nay, that the earth itself is a vale of tears, and the heaven of the Hebrew fanatic our proper home.

"The monastic recluse," says Ulric Hutten, "closes every aperture of his narrow cell on his return from midnight prayers, for fear that the nightingale's song might intrude upon his devotions, or the morning wind visit him with the fragrance and the greeting of the hill forests, and divert his mind to earthly things from things spiritual. He dreads a devil wherever the Nature-loving Greeks worshiped a God." These narrow cells, the dungeons of the Inquisition, the churches whose painted windows excluded not only the air but the very light of heaven, the prison-like convent-schools, and the general control exercised by the Christian priests over the domestic life of their parishioners, laid the foundation of a habit which, like everything unhealthy, became a second nature in old *habitúes*, and gave birth to that brood of absurd chimeras which, under the name of "salutary precautions," inspire us with fear of the night air, of "cold draughts," of morning dews, and of March winds.

I have often thought that *mistrust in our instincts* would be the most appropriate word for a root of evil which has produced a more plentiful crop of misery in modern times than all the sensual excesses and ferocious passions of our forefathers taken together. What a dismal ignorance of the symbolic language by which Nature expresses her will is implied by the idea that the sweet breath of the summer night which addresses itself to our senses like a blessing from heaven could be injurious! Yet nine out of ten guests in an overheated ball-room or travelers in a crowded stage-coach will protest if one of their number ventures to open a window after sundown, no matter how glorious the night or how oppressive the effluvia of the closed apartment. Pious men they may be, and most anxious to distinguish good from evil, but they never suspect that God's revelations are written in another language than that of the Hebrew dogmatist. Here, as elsewhere, men suppress their instincts instead of their artificial cravings. If we have learned to interpret the fact that a child whose mind is not yet biased by any hearsays is sure to prefer pure and cold air to the miasmatic "comfort" of a close room, the troglodyte-habit will disappear, as intemperance will vanish if we recognize the significance of that other fact—that to every beginner the taste of alcohol is repulsive, and that only the tenth or twelfth *dosis* of the obnoxious substance begins to be relished; just as the Russian stage-conductor relishes the atmosphere of his ambulant dungeon, whatever may have been his feelings of horror on the first trip.

If ever we recognize a truth which was familiar enough to the ancients, but seems to have been forgotten for the last ten or twelve centuries, viz., that our noses were given us for some practical purpose, the architecture of our dwellings, our factories, school-rooms, and places of worship, will be speedily corrected; and even the builder of an immigrant-ship will find a way to modify that floating Black Hole of Calcutta called the steerage. Prisons, too, will be modeled after another plan. Our right to diet our criminals on the ineffable mixture of odors which they are now obliged to accept as air depends on the settlement of the question whether the object of punishment is reform or revenge? In the latter case the means answer the purpose with a vengeance indeed: in the first case there is no more excuse for saturating the lungs of a prisoner with the seeds of tuberculosis than there would be for feeding him on trichinæ or inoculating him with the leprosy-virus.

In such countries as Italy and Mexico, where the plurality of the population pass the daylight hours in open air, unventilated bedrooms are almost the only cause of tubercular diseases; but in the north, where children have to be nursed like exotic birds, the chief defects of our domestic arrangements may be classed under three heads: impure air, want of sunshine, and want of room for exercise. The *beau-idéal* of a healthy house would be a well-plastered stone building on some eminence, remote from swamps and stagnant creeks, but surrounded by sunny slopes available for playgrounds; spring or well water; out-door cellar, kitchen in an out-house, or at least not directly below the sitting and sleeping rooms; high ceilings, wainscots, or wall-paper of innocuous colors; deep windows, with projecting mullions to admit the air and exclude the rain; an airy veranda, and no shade-trees on the east and west side, as sunlight is most needed in the mornings and evenings. Children can not thrive in dark back rooms, and in the first eight years of their lives should have all the exercise they want. The countrymen of Dr. Fröbel are ahead in this respect, and the best-arranged nursery I ever saw was the *Findel-zimmer* ("foundling-ward") in the convent of the Ursuline nuns near Würzburg, Germany. The landed estate of the convent having been sequestrated, their department of charitable institutions had been reorganized on a more economical basis, and the poor nuns thought it necessary to apologize for the ingenious simplicity of their Zimmer, whose plan had been suggested chiefly by the necessity of dispensing with hired help. The room was about forty feet square, facing south and west, with three large windows on each side. These windows and the fire-place were barred with net screens, soft to the touch, but securely fastened, and strong enough to stop anything from a foot-ball to a forty-pound baby. The floor was carpeted with rugs, covered with a sort of coarse sheeting to prevent dust. From the floor to the height of the window-sills the walls were padded all round with old blankets, secured with muffled nails, and stuffed with something that felt like moss or cow's hair. The only piece of furniture was a cushioned divan in the corner next to the fire-place; but the floor was covered with playthings and movable nondescripts, balls of all sizes, and a big *Walze*, a sort of wooden cylinder, muffled up with quilts and cotton. From the center of the ceiling depended a hand-swing, two rings just low enough to be within reach of a youngster standing on tiptoe, the original sitting swing having been removed as liable to be used as a catapult in a general row. Above the windows, out of reach of the boldest climber, were shelves with flower-pots, reseda, gillyflowers, and wintergreen. In this in-door Kindergarten, fourteen playmates—twelve babies, namely, and two puppies—had been turned loose, and seemed to celebrate existence as a perpetual circus-game. They could run races, pelt each other with cotton balls, swing in a circle, roll on the floor, and ride the Walze; but the attempt to hurt themselves would have baffled their combined ingenuity. There were no nurslings, of course, but all mischief-ages from three to eleven, wrestling and quarreling now and then, but, as the nuns solemnly averred, never crying except for causes that would make the puppies cry—a squeeze or an inadvertent kick—all disputes being referred to the umpire, a flaxen-haired girl of eight, who often took charge of the Zimmer from morning till night.

The squalling of new-born children can not be helped; puppies will whine, and young monkeys whimper for the first three or four days—it is the novelty of existence that bewilders them—but, if babies of two or three years scream violently for hours together, it generally means that there is something wrong about the management. Indian babies never cry; they are neither swaddled nor cradled, but crawl around freely, and sleep in the dry grass or on the fur-covered floor of the wigwam. Continual rocking would make the toughest sailor sea-sick. Tight swaddling is downright torture; it would try the patience of a Stoic to keep all his limbs in a constrained position for such a length of time; a young ape subjected to the same treatment would scream from morning till night. Forty per cent of all children born in certain manufacturing districts of Belgium and Great Britain die before the end of the second year. They are swaddled, of course; they must not crawl around, and bother people; and "paregoric" does the rest: the child cries for liberty, and receives death. Opiates are sold under right pleasant names nowadays, and at popular prices in the larger cities; but a spoonful of arsenic would be a shorter and a kinder remedy.

Not every family has room and the means to construct a model nursery, but the poorest could spare a few square feet of space in some sunny corner, and, with old quilts and rugs, make it baby-proof enough for all probable emergencies. Then furnish a few playthings and trust the rest to nature. Man wants but little here below, and between meals a pickaninny will content itself with liberty, light and air, and a couple of rag-babies. As soon as a child begins to toddle, it should also have an opportunity to exercise its arms—a grapple-swing, or (if your ceiling be inviolate) a rope stretched from wall to wall. It is surprising how fast the clumsiest youngster begins to profit by such a chance. To the young son of man climbing comes natural enough to shock a witness of anti-Darwinian proclivities. The development of the shoulder-muscles also tends to invigorate the chest, and a fifty-cent hand-swing may save many dollars' worth of cough-medicine.

The progressive development of the motory organs prompts their frequent exercise, and there is no doubt that the gratification of this instinct constitutes the chief element of that physical beatitude which makes the age of childhood the spring-time of every life; and it is equally certain that compulsive physical inactivity inflicts on a healthy child an amount of wretchedness which no prospective advantages can possibly repay. It is hard enough that so large a portion of the human race have to rear their young in a latitude which half the year confines them to the freedom of their four walls; but it is harder that even this limited freedom should be curtailed by so many unnecessary restraints. I wish every houseful of children had a rough-and-tumble room, some out-of-the-way place where the cadets could romp, roll, and jump to their hearts' content. It need not be a heated room nor even an in-door place, as long as it has anything like a roof to it; children are naturally hardy, as they are naturally truthful: effeminacy and hypocrisy are twin daughters of our pious civilization. A wood-shed will do, or a lumber-room with old mattresses and hiding-places. Well-to-do parents might add some gymnastic apparatus, and for big boys a carpenter's table with an assortment of tools; mechanical dexterity may prove useful in many ways, and every normal boy has something of that instinct which phrenologists call *constructiveness*, and which makes the use of such implements a pleasure rather than a task. But, for the youngsters, the rough-and-tumble play is the main thing; it will strengthen their limbs, lungs, and livers, and prevent more ailments than all the pills in Herrick's list of patent medicines. Moreover, it will keep them quiet where other children are sure to be fidgety—in the parlor and at school. Every school-teacher knows that young ruralists are more sedate than city boys; out-door work has given them all the exercise they need; they can take it easy while their comrades are fretting under an irksome restraint. After an hour or two of German gymnastics, combined with wood-chopping and water-carrying, if you like, the wildest boy will prefer a chair to a flying trapeze; for, if the tonic development of the organism is not grossly neglected, sedentary employments *per se* are by no means contrary to nature; in the intervals of their play, the young of frolicsome animals will sit motionless for hours; even kittens and young monkeys; not to mention colts which have off-days, when they won't stir a foot if they can help it.

It would be a great improvement on our present system of school-education, if children could learn the rudiments at home and pass their infancy, the first eight or ten years, at least, under the immediate supervision of their parents; a transition-period of three or four years of home studies would help them to steer clear of many moral and physiological cliffs. It is always the best preparatory school; only a private teacher has time and patience to *interest* a pupil in the dry *principia* of every science; but a still greater advantage is his independence of fixed methods and fixed hours. As a general rule, the forenoon is the best time for studies, and the airiest room in the house the best locality. Pure air has a wonderful effect on the clearness of our cerebral functions; the half-suffocating atmosphere of the average school-room is as stupefying as the influence of a half-intoxicating drink. Heat aggravates the offensiveness of foul air; but in a well-ventilated room the degree of temperature is comparatively unimportant. As it would be inconvenient to load ourselves with blankets in day-time, less than 50° Fahr. would make sedentary occupations rather uncomfortable, and more than 80° would become oppressive in a close apartment; but between these extremes we may safely suit our convenience. Perfectly pure or perfumed air may be very warm and still very pleasant, as all know who have entered a conservatory or a tidy baker's shop on a cold winter day.

In large town schools, where hundreds of children have to breathe the same air, I would advise a change of rooms from hour to hour, and a thorough renovation of the vitiated atmosphere by opening every window and every door, and keeping up a rousing fire. The air-currents could be re-enforced by mechanical means —canvas-floppers or revolving fans—and *fumigation* would greatly aid the good work. The South European druggists sell various kinds of frankincense that can be burned on a pan or a common stove, and will fill a large church with odors more or less Sabæan, according to price—ten cents' worth a day would be enough to beatify a whole town school; Mohammed, the man of God, included perfume among the three greatest blessings of human life. Young children ought to have a recess after every lesson, and should not be required to sit rigidly quiet. The best writing-stand for children is Schreber's "telescope-desk," a box-like contrivance, with a movable top that can be lowered or raised to suit the convenience of sitting or standing writers. In a latitude where the weather so often precludes the possibility of out-door recreations, every school-house should have a recess-room, and every town school an in-door gymnasium.

Fireside comforts are almost inseparable from the idea of an open fire-place, and from an hygienic standpoint, too, the old-fashioned chimney, or an open grate, is far superior to a closed stove. But it should not be forgotten that the operation of the chimney-draught alone is insufficient to correct the vitiated air of a small room, it merely creates an outward current. An open window completes the renovating process; in cold weather a few minutes are sufficient to revitalize the in-door atmosphere for a couple of hours. Only the blindest prejudice can deny the pleasant effect of such an influx of life-air; it revives the azotized lungs as a draught of cool water refreshes the parched palate. Colds are never taken in *that* way. The very name is a misleading misnomer—*infection* or influenza would be the right word. Long exposure to a freezing storm, in certain cases, induces a true pleuritic fever, a very rare affection, and entirely different from the only too familiar catarrh. What we call a *cold* (*refroidissement, Erkältung*) is caused by the influence of impure air, or dust, on the sensitive tissue of our respiratory organs; subsequent exposure to the open air merely initiates the crisis of the disorder, the discharge of the accumulated mucus through the nose or throat. Fresh air is here only the proximate cause, as in toothache, or in those paroxysms of retching following upon the first respiration of a half-drowned person. If we postpone the crisis by persistently avoiding the open air, the unrespirable matter, instead of being discharged, will be deposited in the tissue of the lungs in the form of tubercles.

In the chapter on Diet I have stated the physiological objections to a late supper, and I will here mention an additional reason why the afternoon meal should be the last: It would give an overworked mother a chance to close the kitchen-door at six o'clock, and devote the rest of the evening to her family. Domestic habits depend greatly upon the employment of the long winter evenings that have to be passed in-doors somewhere; whether at home or—elsewhere, depends upon home-comforts rather than upon home-missions; a treatise on the art of making the chimney-corner attractive would be the most effective temperance lecture. Fredrika Bremer recommends fairy stories; in a North American city Scheherezade would probably avail herself of the circulating library, and a fascinating story-book is, indeed, an excellent substitute for the old-fashioned remedies against gadding. Good books, flowers, and music, combined with pleasant conversation and a cheerful fire, would neutralize the attractions of the average " saloon." Playthings and social games, too, would help to compensate the youngsters for the want of out-of-door sports, and where they have a room to themselves I would suggest the introduction of some entertaining pet, a raccoon or a tame squirrel-monkey. Let the boys have some fun—provide pastimes; it is *ennui* rather than natural perversity that leads our young men to the rum-shop.

The end of the day is the best time for a sponge-bath; a sponge and a coarse towel have often cured insomnia where diacodium failed. A bucketful of tepid water will do for ordinary purposes; daily cold shower-baths in winter-time are as preposterous as hot drinks in the dog-days. Russian baths and ice-water cures owe their repute to the same popular delusion that ascribes miraculous virtues to nauseating drugs—

the mistrust of our natural instincts, culminating in the idea that all natural things must be injurious to man, and that the efficacy of a remedy depends on the degree of its repulsiveness. Ninety-nine boys in a hundred would rather take the bitterest medicine than a cold bath in mid-winter. If we leave children and animals to the guidance of their instincts they will become amphibious in the dog-days, and quench their thirst at the coldest spring without fear of injurious consequences; but in winter-time even wild beasts avoid immersion with an instinctive dread. A Canadian bear will make a wide circuit, or pick his way over the floes, rather than swim a lake in cold weather. Baptist missionaries do not report many revivals before June. Warm springs, on the other hand, attract all the birds and beasts that stay with us in winter-time; the hot spas of Rockport, Arkansas, are visited nightly by raccoons and foxes in spite of all torch-light hunts; and Haxthausen tells us that in hard winters the thermæ of Pactigorsk, in the eastern Caucasus, attract deer and wild-hogs from the distant Terek Valley. I know the claims of the hydropathic school, and the arguments *pro* and *con*, but the main points of the controversy still hinge upon the issue between Nature's testimony and Dr. Priessnitz's.

Our beds are our night-clothes, and ought to be kept as clean as our shirts and coats. Woolen blankets are healthier than quilts; put a heavy United States army blanket over a kettle full of hot water and see how fast the steam makes its way through the weft; a quilt would stop it like an iron lid, and thus tends to check the exhalation of the human body. In order to disinfect a quilt you have first to loosen the pressed cotton; a woolen blanket can be steamed and dried in a couple of hours. For similar reasons a straw tick is better than a horse-hair mattress, though a woven-wire mattress is perhaps preferable to both. Feather-beds are a recognized nuisance. Children over ten years should sleep alone, or at least under separate blankets, if the bedsteads do not reach around.

If you would preserve your children from wasting diseases, do not stint them in their sleep; chlorotic girls, especially, and weakly babies need all the rest they can get. If they are drowsy in the morning, let them sleep; it will do them more good than stimulants and tonic sirups. For school-children in their teens, eight hours of quiet sleep is generally enough, but do not restrict them to fixed hours; in midsummer there should be a *siesta-*corner in every house, a lounge or an old mattress in the coolest nook of the hall, or a hammock in the shade of the porch, where the little ones can pass the sleep-inviting afternoons. Nor is it necessary to send them to bed at the very time when all nature awakens from the torpid influence of the day-star; sleep in the atmosphere of a stifling bedroom would bring no rest and no pleasant dreams. But an hour after sunset there will be a change; the night-wind rises and the fainting land revives; cool air is a febrifuge and Nature's remedy for the dyspeptic influences of a sultry day. Open every window, and let your children share the luxury of the last evening hour; after breathing the fresh night-air for a while they will sleep in peace.

CHAPTER III.

OUT-DOOR LIFE.

"Disease is a hot-house plant."—HALLER.

EVERY disease is a protest of Nature against an active or passive violation of her laws. But that protest follows rarely upon a first transgression, never upon trifles; and life-long sufferings—the effects of an incurable injury excepted—generally imply that the sufferer's mode of life is habitually unnatural in more than one respect. For there is such a thing as vicarious atonement in pathology: a strict observance of any one of the three or four principal health-laws rarely fails to reward itself by a long immunity from the consequences of otherwise evil habits. Frugality thus counteracts the morbific tendency of indolence; perfect continence may steel even a feeble constitution against the effects of hunger and overwork; and, by avoiding the great vice of intemperance, the Epicureans atoned for a multitude of minor sins.

But the surest of all natural prophylactics is active exercise in the open air. Air is a part of our daily food and by far the most important part. A man can live on seven meals a week, and survive the warmest summer day with seven draughts of fresh water, but his supply of gaseous nourishment has to be renewed at least fourteen thousand times in the twenty-four hours. Every breath we draw is a draught of fresh oxygen, every emission of breath is an evacuation of gaseous recrements. The purity of our blood depends chiefly on the purity of the air we breathe, for in the laboratory of the lungs the atmospheric air is brought into contact at each respiration with the fluids of the venous and arterial systems, which absorb it and circulate it through the whole body; in other words, if a man breathes the vitiated atmosphere of a factory all day and of a close bedroom all night, his life-blood is tainted fourteen thousand times in the course of the twenty-four hours with foul vapors, dust, and noxious exhalations. We need not wonder, then, that ill-ventilated dwellings aggravate the evils of so many diseases, nor that pure air should be almost a panacea.

Out-door life is both a remedy and a preventive of all known disorders of the respiratory organs; consumption, in all but the last stage of the *deliquium*, can be conquered by transferring the battle-ground from the sick-room to the wilderness of the next mountain-range. Asthma, catarrh, and tubercular phthisis, are unknown among the nomads of the intertropical deserts, as well as among the homeless hunters of our Northwestern Territories. Hunters and herders, who breathe the pure air of the South American pampas, subsist for years on a diet that would endanger the life of a city dweller in a single month. It has been repeatedly observed that individuals who attained to an extreme old age were generally poor peasants whose avocations required daily labor in the open air, though their habits differed in almost every other respect; also that the average duration of life in various countries of the Old World depends not so much on climatic peculiarities or their respective degree of culture as on the chief occupation of the inhabitants; the starved Hindoo outlives the well-fed Parsee merchant, the unkempt Bulgarian enjoys an average longevity of forty-two years to the west Austrian citizen's thirty-five.

In the cities of the higher latitudes, sedentary occupations in a vitiated atmosphere become often a sort of "second nature": artisans and shop-keepers, after following their business for a number of years, frequently come to dislike fresh air, as the convent slave, by an analogous suppression of his better instincts, becomes averse to free inquiry. But this abnormal indolence seldom becomes hereditary—perhaps never, if we except the children of inebriate idiots. The mediæval prejudice against all natural propensities—founded on the dogma of innate depravity—is, indeed, strikingly refuted by a young child's love of out-door exercise. Without the mediation of supernatural revelators or preternatural bugbears, a healthy boy prefers even the hardships of our northern winter sports to the atmosphere of a comfortable stove-room, and in summer-time the paradise of childhood is still a tree-garden. No domestic events of our later years can efface the impression of the woodland rambles, butterfly hunts, and huckleberry expeditions of our boyhood: the recollections of our first out-door adventures endure like the mountains and rivers of a promised land whose cities have vanished for ever.

I have often been asked at what age infants can first be safely exposed to the influence of the open air. My answer is, On the first warm, dry day. There is no reason why a new-born child should not sleep as soundly under the canopy of a garden-tree on a pillow of sun-warmed hay as in the atmosphere of an ill-ventilated nursery. Thousands of sickly nurslings, pining away in the slums of our manufacturing towns, might be saved by an occasional *sun-bath*. Aside from its warmth and its chemical influence on vegetal oxygen, sunlight exercises upon certain organisms a vitalizing influence which science has not yet quite explained, but whose effect is illustrated by the contrast between the weeds of a shady grove and those of the sunlit fields, between the rank grass of a deep valley and the aromatic herbage of a mountain meadow, as well as by the peculiar wholesome appearance of a "sunburned" person or a sun-ripened fruit. Sunlight is too cheap to become a fashionable remedy, but its hygienic influence can hardly be overrated. Even in the glorious climate of the Latian hills, the Roman Epicureans constructed special *solaria*—glass-covered turrets—where they could bask in the full rays of the winter sun, the balm of old age, as Columella calls it; and, on the summerless Isle of Rügen, Nature has taught the poor fishermen to carry their bairns to the downs of Stubbenkammer, whenever the Baltic fogs alternate with a few sunny days. Dry sand is, indeed, an excellent medium of solar caloric. Children like it instinctively; most babies are fond of rummaging in some tangible, yielding element. In default of a sunny beach, get a car-load of river-sand, spread it and expose it to the sun for a couple of hours, then rake it together, mix it *ad captandum* with a bushel of pebbles (good-sized ones, lest they might be mistaken for sugar-plums), divest your *bambino* of all superfluous clothing, and let him wallow —all afternoon, if he chooses; if the surface of the pile gets too warm, instinct will teach him to dig down to the cooler substrata. Or take him to a meadow where fresh hay has been piled up in little stacks; climbing and tumbling will do him more good than lying motionless in a narrow baby-carriage. The inventor of the Kindergarten recommends a grassy hollow with scattered playthings, piles of dry leaves, etc. (near a shade-tree in midsummer), where young squealers can take care of themselves for an hour or two, and warrants that they will not cry, unless their botanic researches should happen to acquaint them with the properties of the German horse-nettle. On mild winter days, too, self-motive babies ought to pass a few hours out-of-doors, even if the ground be a little wet; a sunny nook on the lee-side of a garden-wall is a healthier playground than the dusty floor of a stove-room.

From the fourth to the end of the fourteenth year children should spend the larger part of every summer in out-door exercises. Next to a total reform of our dietetic habits, a general observance of this rule would be the surest way to regain the hardiness and longevity of our forefathers. The years of growth lay the foundation of our bodily constitution, and, under favorable circumstances, the human system, during that period, seems to accumulate a surplus of physical vigor, which in after-life will become available as an annuity-fund of health and happiness. Education, like charity, ought to begin at home; in boarding-colleges, protectories, orphan asylums, etc., the rudiments should be taught in *winter schools*. At the price of life-long infirmities precocious erudition is too dear-bought; besides, it should not be forgotten that in the years when students can take a personal interest in their lessons they will make more progress in a single month than during years of involuntary confinement in boy-pens, as Dr. Salzmann calls our municipal baby-schools. The employment of young children in cotton-factories is a crime against society, and ought to be legally prohibited, like the trade in Italian organ-boys and Chinese slave-girls. Swiss artisans, who have passed their boyhood in the mountains, are comparatively proof against the influence of in-door occupations. And, in the mean time, out-door life need not be a life of idleness. That children are fond of play means simply that they prefer entertaining employments to tedious ones. Youngsters under five years gambol instinctively like young puppies, in order to acquire the art of locomotion, but soon afterward they begin to play with a conscious purpose, and do not object to playing at something profitable; young savages and peasant-boys join in the labors of their parents with an eagerness that vindicates human nature against the charge of innate frivolity. Make your boy a Jack-of-all-out-door-trades before you make him a classic polyglot, and, if you destine him for any trade in special, let him play with the tools of that special trade. "The best plan of education," says Goethe, "is that of the Hydriotes, the Greek trading-sailors, who take their infant boys out to sea and let them sport around amid oakum and belaying-pins before they learn to handle them with a business purpose. Such a school has graduated the heroes who with their own hands could grapple the fire-boat to the flag-ship of the enemy."

Even for their children's sake, married men should never quarter their families in the heart of a great city. Not everybody can own a farm, but, wherever the suburban cottages adjoin waste building-lots and dry ravines, there will be no lack of opportunities for out-door pastimes. Let the girls make weed-brooms, and the boys construct fortifications, *à la* Uncle Toby, if they can do no better, and miss no chance to send them out in the country for a day or two. Our town parks are too exclusive; sauntering between inviolate grass-plots and prohibitory placards is dull work for urchins that long to commit horse-play; but there are few cities, even on the Atlantic sea-board, where the "open country"—woods, fallow fields, and hill-sides—could not be reached by a two hours' walk. There let your children spend every sunny afternoon; make arrangements with your neighbors, and engage a guide if you can not afford to go yourself; teach the youngsters to collect beetles and butterflies, encourage the fern mania if your girl has outgrown the buttercup period, connive at a bird's nest or two, do anything to keep them out of the tenement dungeons. If you are blessed with a farm (or a tolerant country cousin), hay-making, apple-gathering, turkey-herding, repairing of ditches and garden-walls, will make earth an Elysium to every normal child; never mind the weather; a summer shower, a chilly morning, or a hot afternoon will not hurt a healthy boy, and the girls will take care of themselves—or rather of their dress—if the grass is wet. If you send them to school before their teens, give them at least the full benefit of their vacations and of every free Saturday. In fall and winter a day of athletic field-sports will keep a boy in tolerable health for the rest of the week, and a vacation tour of six or eight weeks may atone for many months of sedentary life.

In the preceding chapter I have pointed out the main cause of catarrhal affections. With the exception of deep-seated breast-coughs, "colds" may be nipped in the bud by a few hours of hard, *sudorific* work in the open air. It may be an heroic cure, requiring a good deal of will-force in cold weather, but it is an infallible and the only radical remedy. In half a day the nasal ducts and the perspiratory exhalants will throw off irritating matters which would defy the drug-doctor for a couple of weeks, or yield only to exercise their influence in another direction, for poison-remedies merely change the form of a disease. But the beneficial effect of out-door exercise is not limited to the respiratory organs: their quickened function reacts on the digestive apparatus, on the nervous system, and through the nerves on the mind; true mental and physical vigor in any form can be maintained only on a liberal allowance of life-air; those who feed their lungs on miasma become strangers to that exuberant health which makes bare existence a luxury. After years of in-door life, the victims of melancholy, dyspepsia, and dull headaches come to accept their discomforts as the normal condition of mankind, but upon the first appearance of such disorders our instinct suggests the cause and the cure with an urgency which makes confinement in the atmosphere of our northern dwelling-houses the greatest affliction of childhood. If we reflect on the fact that our earth is surrounded by a respirable atmosphere of at least eight hundred million cubic miles, it seems a sad comment on the enlightenment of modern civilization that the unsatisfied thirst after life-air should inflict more misery upon millions of our fellow-men than hunger and all the hardships of poverty combined. " On the day of judgment," says Jean Paul, " God will perhaps pardon you for starving your children when bread was so dear; but, if he should charge you with *stinting them in his free air,* what answer shall you make?"

Perfect health depends upon a daily supply of fresh air as much as on our daily bread; but within certain limits the human organism is capable of adapting itself to abnormal circumstances. A man may accustom himself to devour his weekly allowance of solid food at a single meal, and in a similar way the vitalizing elements of air and sunshine can be hoarded up—allotropically, for all we know—for days, weeks, and months in advance. The Zooloo hunter who, after a six days' fast, gets a chance to satisfy the cravings of his stomach, can not be expected to content himself with half-pint rations *à la* Luigi Cornaro, and in midsummer, after six months of sedentary life, a boy should get his fill of out-door exercises; let him drink sunlight at every pore, do not stint his allowance of oxygen, compensate him for very long arrears of woodland air and mountain-rambles.

With a little experience vacation trips can be managed very cheaply. Professor Jordan, of the Ilefeld *Pedagogium*, takes his summer boarders to the Hartz, or even to the Austrian Alps, at an aggregate daily expense of fifteen marks (three and a half dollars) for twenty or twenty-five big boys with North-German appetites. They carry their own beds in the form of a plaid and a plair of foot-sacks (boot-like felt socks), and sleep wherever they find a shade-tree or an open barn. Their portable commissariat consists of biscuits and brown sugar; with fresh milk and such *entremets* as the mountain inns may afford, they make out two good meals a day, besides occasional luncheons of nuts and huckleberries. Twenty-two of the twenty-four hours are thus spent in the open air, but the long summer days are almost too short for all the entertainments on the liberal professor's programme. Zoölogy, botany, and geology are only collateral pursuits, the main thing is the uproarious fun in the mountains; climbing cliffs, tumbling bowlders from projecting rocks, and chasing squirrels from tree to tree do not endanger the toilet of the excursionists, for every one of them wears *turnerdrell*, a sort of coarse linen, as tough, though not quite as soft, as corduroy.

Observant managers of such expeditions soon get rid of the dismal prejudices against cold spring-water, "wet feet," and "untimely baths." The craving of a thirsty wanderer after cold water is not an abnormal appetency, but a natural instinct, and can be indulged with perfect impunity; a bath in sun-warmed river-water is healthy as long as it is enjoyable; South-Sea Islanders and the children of the Genoese fishermen spend whole afternoons in the surf, and—barring sharks and medusas—without fear of dangerous consequences. There is no harm in wet stockings as long as the feet are in motion; at home it is perhaps better to change them at once, though the Canadian lumbermen dry them on their legs before the camp-fire, or even in bed —i. e., under a pair of "Mackinaw blankets," which blankets have often served as overcoats during the day, but in the course of the night are dried by the animal warmth like a pack of wet sheets. Sun-strokes can be obviated by a simple and very inexpensive precaution —temporary abstinence from animal food. A refrigerating diet (vegetables, fruit, etc.) counteracts the effect of a high atmospheric temperature, but the calorific influence of meat and fat, combined with solar heat and bodily exertion, overcomes the organic power of resistance, the pyretic blood-changes produce congestion of the brain and sometimes instant death. I venture the assertion that in nineteen out of twenty cases of comatose sun-stroke it will be found that the victims were persons who had gone to work in the hot sun after a meal of greasy viands. One to two P. M. is the sun-stroke-hour.

Among the permanent benefits which young persons may derive from a pedestrian tour, it is not the least that they will mostly get rid of the night-air superstition. Sweet rest and pleasant dreams he knows not who has never slept under a Mexican live-oak tree on a bundle of fresh-plucked Spanish-moss, or in the loft of a Tennessee cotton-gin while the winds of the summer night play in draughts and counter-draughts through four open louvres. The advantages of a hardy education in all such things are quite incalculable; the word *hardiness* sums up the chief characteristics that distinguished the moral and physical life of the ante-Christian ages from the scrofulous effeminacy of our stove-room civilization.

The teachers of the *Pedagogium* and similar institutions assured me that their scholars were never more *aufgeweckt* (wide-awake) than during the first six or eight weeks after the long vacations; even the drawing-masters had no reason to complain about "club-fists." It is a very common but quite erroneous notion that the burly strength of the human hand impairs its capacity for delicate manipulations: the iron-fisted Gemsenjäger of the Tyrolese Alps are the nicest marksmen; and Leonardo da Vinci, who could draw a perfect circle without a compass, could not the less break a silver piaster between his two thumbs and two forefingers.

The Ilefelders were also the first to make Saturday an hygienic sabbath. In spring and fall, all such Saturdays should be consecrate to the wood-gods; leaf-forests, under the influence of sunlight, exhale the antidote of our atmospheric poisons. Start the youngsters at sunrise with a basketful of cold meats, and orders for an equal quantity of strawberries, or, if the woods are safe, let them go on Friday night, and camp in the open air; they will long for the advent of that night as Tom-a-lin for the festival of the fairies. Let them rise with the sun and spend the whole day in active exercise, the merrier the better; in a mountain country arrange a new programme for every week, explore the local Ararats, and let the boys scale them in succession, as the members of the Alpine Club tackle their bergs and horns. If the weather should disappoint you, do not hesitate to *improve* the next sunny day, though it should happen to be a Sunday. The God of Nature can be worshiped in his own temple: the wonder of his living world is his most authentic revelation. Where Sunday is the only free day in the week, no puritanical tyranny or Jesuitical ingenuity will ever prevent the poor from making it a day of recreation; the only question is, whether that recreation shall be sought in the secret rum-shops and back-alleys of the city, whose gates the sabbatarians would shut upon us, or in the free woods and mountains, where the worshiper of the All-Father can find inspiration as well as joy and health. The wood-thrush, it is true, does not modulate her anthems in a whining drawl; the pine-tree lifts his head without fear of provoking his Creator by a want of crawling humility; no dread of a joy-hating priest-god disturbs the gambols of the squirrel and the aërial dances of the brook-midge; the butterfly and the humming-bird do not think it necessary to "mortify the eye with dreary drab," but their happiness imparts a lesson not less divine for being at variance with the doctrines of an atrabilious fanatic.

According to the Grecian allegory, the wood-craft goddess Diana was the antagonist of the Cyprian Venus; and a *penchant* for out-door sports is indeed the best safeguard against certain vices of youth. The precocious Don Juans of our great cities could be more easily reformed by a hunting expedition to the next Sierra Nevada than by all the homilies of Fray Gerundio. Like depraved humors, prurient propensities yield to active exercise more readily than to physic and prayer. Hunting tribes are generally continent, stalwart, and comely; wood air is a cosmetic; the finest types of the human form are not found within the precincts of the Palais Royal, but in the Caucasus and the Kentucky forest counties.

Enjoyable winter excursions are a privilege of the rich; still, a pair of good skates make a convenient pond or a small river a great blessing. From a sanitary point of view, the neighborhood of larger streams is not so much of an advantage; besides being the terror of parents during the skating season, a big river is apt to render the contiguous lowlands more or less malarious, especially after every inundation. In snow-bound villages children have to depend mainly on in-door exercises; cold air, however, is a powerful tonic, and a two hours' snow-ball fight will generally suffice to vitalize a juvenile constitution for a couple of days. Mountain air, too, is a peptic stimulant, and pedestrian excursions are doubly invigorating if they include a good deal of up-hill work.

For those who wish to select their dwelling-place with regard to the hygienic interest of their children, the best location is, therefore, on the whole, the bank of a small river in the neighborhood of a large mountain-range.

CHAPTER IV.

GYMNASTICS.

"Force begets Fortitude and conquers Fortune."—HELVETIUS.

PHYSICAL vigor is the basis of all moral and bodily welfare, and a chief condition of permanent health. Like manly strength and female purity, gymnastics and temperance should go hand in hand. An effeminate man is half sick; without the stimulus of physical exercise, the complex organism of the human body is liable to disorders which abstinence and chastity can only partly counteract. By increasing the action of the circulatory system, athletic sports promote the elimination of effete matter and quicken all the vital processes till languor and dyspepsia disappear like rust from a busy plowshare. "When I reflect on the immunity of hard-working people from the effects of wrong and overfeeding," says Dr. Boerhaave, "I can not help thinking that most of our fashionable diseases might be cured *mechanically instead of chemically*, by climbing a bitterwood-tree or chopping it down, if you like, rather than swallowing a decoction of its disgusting leaves." The medical philosopher, Asclepiades, Pliny tells us, had found that health could be preserved, and, if lost, restored, by physical exercise alone, and not only discarded the use of internal remedies, but made a public declaration that he would forfeit all claim to the title of a physician if he should ever fall sick or die but by violence or extreme old age. Asclepiades kept his word, for he lived upward of a century, and died from the effects of an accident. He used to prescribe a course of gymnastics for every form of bodily ailment, and the same physic might be successfully applied to certain moral disorders, incontinence, for instance, and the incipient stages of the alcohol-habit. It would be a remedy *ad principium*, curing the symptoms by removing the cause, for some of the besetting vices of youth can with certainty be ascribed to an excess of that potential energy which finds no outlet in the functions of our sedentary mode of life. In large cities parents owe their children a provision for a frequent opportunity of active exercise, as they owe them an antiseptic diet in a malarious climate.

Nor can this obligation be evaded by depreciating the importance of physical culture as distinct from that of the mental faculties. For the term of their earthly pilgrimage the human body and the most immortal soul are more inseparable and more interdependent than the horse and its rider: a Centaur would hardly have promoted his higher interests by neglecting the equine part of his person. "I have sinned against my brother, the Ass," said St. Francis, when the abuse of his body had brought on a mortal disease. For the idea that the supremacy of the mind could be enforced by debilitating penances is a fatal mistake; an enervated body, instead of ministering to the needs of the mind, becomes its tyrant, a querulous, capricious, and exorbitant master. Every hospital attendant knows that, with the rarest exceptions, the sufferers from exhausting diseases have no more self-control than a fretful child. Neither can the progress of our mechanical industries be made a pretext for undervaluing the advantages of an athletic education. It has been prophesied that the time will come when the autocrat of the breakfast-table shall break his egg with a dynamite wafer; but, unless we invent a labor-saving contrivance for every muscle of the human organism, there is not a day in the year nor an hour in the day when the practical business of life can not be performed more easily and more pleasantly with the aid of a vigorous body, not to remention the moral disadvantages which never fail to attend the loss of manly self-reliance. Active exercises also confer beauty of form and a natural grace of deportment. "By their system of physical culture," says a Scotch author, "the Greeks realized that beautiful symmetry of shape which for us exists only in the ideal, or in the forms of divinity which they sculptured from figures of such perfect proportions."

That a man's welfare in every sense of the word depends upon the normal development of his body might, therefore, seem an axiom whose self-evidence could be questioned only in a fit of insane infatuation; yet an Oriental fanatic has succeeded in tainting countless millions of his fellow-men with this very insanity. About six hundred years before the beginning of our chronological era, a speculative philosopher of Northern Hindostan set about to investigate the origin of the sufferings which so often make human life a burden instead of a blessing, and, failing to trace these afflictions to any avoidable cause, he took it into his head that terrestrial existence itself must be an evil, and conceived the unhappy idea of preaching a crusade against the love of earth and the rights of the human body, as distinct from a supposed preternatural part of our being. His success has been, beyond all compare, the greatest calamity that ever befell the human race since the days of the traditional deluge; not only that the doctrines of Gautama bore their fruit in the utter physical degeneration of his native country, and the populous empires of Eastern Asia, but, seven centuries after, the essential doctrines of Buddhism, intensified by an admixture of Gnostic demonism and Hebrew mythology, were preached upon the shores of the Mediterranean and invaded the paradise of the Aryan nations. A mania of self-torture and miracle-worship broke out like a mental epidemic, and, at the very time when the influence of Grecian civilization began to wane, the new creed spread into Italy, and the friends of science and freedom were confronted with the fearful danger of an antinatural religion. What that danger meant, our liberated age can hardly realize unless we review the fate of those nations to whom salvation came too late; on whose destiny the curse of that superstition has been wrought out to the bitter end. The attempt to carry the theories of the Hebrew fanatics into practice led to a state of affairs against which the *unpossessed* part of mankind had to combine in sheer self-defense; the maniacs were overpowered, but only after a struggle which has trampled the chief battle-fields into dust, and not before they had turned the Mediterranean Godgarden into such a pandemonium of madness, tyranny, and wretchedness, that the lot of the African savages appeared heaven in comparison. The annals of pagan despotism furnish no parallel to the pages stained with blood and tears that record the horrors of the inquisitorial butcheries and man-hunts of the middle ages. The history of science is the history of a day with a bright morning and a sunny evening, but interrupted at the noontide hour by a total eclipse of common sense and reason. The men that inculcated a belief in the possibility of witchcraft and demoniac possession are responsible for the agonies of the three million human beings that perished in the flames of the stake; the dogma of total natural depravity guided the arm that aimed its poisoned daggers at the heart of every social, political, or scientific reformer. But the direst of all the evils which made the rule of the **miracle-mongers** the unhappiest period in the history of this earth was, after all, their total neglect of physical education—the logical outcome of their Nature-hating insanity. Their disciples were assured, in the name of an infallible revelator, that all earthly concernments are vain; that we can not please God without mortifying our bodies; that our natural instincts must be suppressed, in order to qualify our souls for the New Jerusalem. The joys of Nature were to be shunned as man-traps of the archfiend. Sickness was to be cured by prayer and certain ecclesiastic ceremonies. "Bodily exercise," we are informed, "profited but little." The Olympic games were suppressed by order of a Christian emperor.* The health-code of the Mosaic dispensation was repealed as unessential, and indeed superfluous, in a community of miracle-workers who could defy the laws of Nature with the aid of supernal spirits. Gluttony and besottedness were encouraged by the example of the ministers of that creed. **Manly exercises, the festivals of the seasons, mirth, pastimes, and health-giving sports were discouraged as unworthy of a true saint**; the sons of the thaumaturgic church were taught that our natural desires and natural dispositions are wholly evil; that the study of worldly sciences is vain, and solicitude for the welfare of the body a proof of an unregenerate heart.

* " A. D." 394.

To these doctrines we owe the consequences of our countless sins against the physical laws of God; the many irretrievable losses by the ruin of a former civilization; the terrible night of the long centuries when science was paralyzed, when industrial progress was limited to the invention of new instruments of torture, when the neglect of husbandry changed so many Elysian fields into hopeless deserts. To these doctrines the Latin peoples owe the sickliness and effeminacy which contrast their present generation with the hero-race of antiquity. It is a favorite subterfuge of the Jesuitical apologists to ascribe that degeneracy to climatic influences. A cold climate has not saved the North-China votaries of Buddhism, and would not have saved the North-Europeans against a prolonged influence of Hebrew Buddhism. We must not forget that in Northern Europe the rule of the anti-naturalists did not begin before the end of the seventh century, and never overcame the latent *protestantism* of the Teuton races. In a warmer country than Italy the votaries of the manlier prophet of El Medina have always preserved their physical vigor, and the representative North-African of the present day is the physical superior of his South-European contemporary, while the forefathers of the same African were mere children in the hands of the palæstra-trained Roman warrior.

The physical corruption of the non-Mohammedan inhabitants of Southern Europe and Southern Asia has reached the incurable stage of complacent effeminacy: their indifference to the vices of indolence precludes the possibility of reform. Indifference to physical degradation is, indeed, a symptom of a deep-seated disease. Mental inertness is often but a dormant state of the intellect, a state from which the sleeper may be roused at any moment by the din of war, by the light of a great discovery, by the voice of an inspired poet. Physical indolence is the torpor which precedes the sleep that knows no waking. The civilization of Greece, Dutch art, the science of Bagdad and Cordova, sprang up, like water from the rock of Moses. Can historians point out a single instance of an unmanned people regaining their manhood? The bodily degeneracy of a whole nation dooms it to a hopeless retrogression in prosperity and political power.

The first use we should make of our regained liberty is, therefore, the re-establishment of those institutions to whose influence the happiest nations of antiquity owed their energy and their physical prowess, their martial and moral heroism, their fortitude in adversity. The physical constitution of man was never intended for the sluggish inactivity of our sedentary and sabbatarian mode of life. In a state of nature, the faculty of voluntary motion distinguishes animals from plants, and our next relatives in the great family of the animal kingdom are the most restlessly active of all warm-blooded creatures. The children of Nature—hunters, shepherds, and nomads—pass their days in out-door labor and out-door sports; physical exercise affords them at once the necessaries of life and the means of recreation, and secures them against all physical ills but wounds and the infirmities of extreme old age. Civilization, i. e., life on the co-operative plan, exempts many individuals from the necessity of supplying their daily wants by daily physical labor; wealth removes the objective necessity of physical exercise, but the subjective necessity remains; millions of city-dwellers, in their pursuit of artificial luxuries, stint their bodies in the natural means of happiness; they increase their stock of creature-comforts and decrease their capacity for enjoying them; religious and social dogmas pervert their natural instincts; their children are crammed with metaphysics till they forget the physical laws of God.

These evils the inventors of gymnastics managed to counteract, and, before we can hope to recover the Grecian earth-paradise, our system of public education needs an essential and thorough reform. On earth, at least, moral and physical culture should be as inseparable as mind and body; every town school should have an indoor and out-door gymnasium; the same village carpenter who takes a contract for a dozen rustic school-benches should get an order for an horizontal bar and a couple of jumping-boards; every school district should appoint a superintendent of gymnastics; every town a committee of public arenas: cities that can afford to devote a hundred tax-free tabernacles to Hebrew mythology might well spare an acre of ground for Grecian athletics.

At a very early period the Greeks of Southern Europe and Asia Minor had recognized the truth that, with the advance of civilization and civilized modes of life, a regular system of bodily training must be substituted for the lost opportunities of physical exercise which Nature affords so abundantly to her children in the daily functions of their wild life. "It is impossible to repress luxury by legislation," says Solon, in Lucian's "Dialogues of Anacharsis," "but its influence may be counteracted by athletic games, which invigorate the body and give a martial character to the amusements of our young men."

The nature of ancient weapons and the use of heavy defensive armor made the development of physical force a subject of national importance, but military efficiency was by no means the exclusive object of gymnastic exercises. The law of Lycurgus provides free training-schools for the thorough physical education of both sexes, and cautions parents against giving their daughters in marriage before they had attained the prescribed degree of proficiency in certain exercises, which were less ornamental and probably less popular than what we call calisthenics. Grecian physicians, too, prescribed a course of athletic sports against various complaints, and had invented a special curriculum of gymnastics, which, as Ælian assures us, never failed to cure obesity. When the increase of wealth and culture threatened to affect the manly vigor of the race, physical education was taken in hand by the municipal authorities of almost every Grecian city; and the ablest statesmen of Athens, Thebes, and Corinth, emulated the Spartan legislator in founding palæstræ, gymnasia, and international race-courses, and devising measures for popularizing these institutions. Four different localities—Olympia, Corinth, Nemea, and the Dionysian race-course near Athens—were consecrated to the "Panhellenic games," at which the athletes of all the Grecian tribes of Europe and Asia met for a trial of strength at intervals varying from six months to four years, the latter being the period of the great Olympic games which formed the basis of ancient chronology. The honor of being crowned in the presence of an assembled nation would alone have sufficed to enlist the competition of all able-bodied men of a glory-loving race, but many additional inducements made the Olympic championship the day-dream of youth and manhood, and served to increase the ardor of gymnastic emulation. The victors of the Isthmian and Nemean games were exempt from taxation, became the idols of their native towns, were secured against the vicissitudes of fortune and the wants of old age, by a liberally-endowed annuity fund, and enjoyed all the advantages and immunities of the privileged classes.

Egenetus, a humble citizen of Agrigentum, won three out of the five prizes of the ninety-second Olympiad, and was at once raised from poverty to opulence by the magnificent presents which the enthusiasm of the spectators forced upon him before he had left the arena. His return to his native city was attended by a procession of three hundred chariots, each drawn, like his own, by two white horses, and all belonging to the citizens of the town. All international quarrels and family feuds were suspended when the preparatory interval of forty-eight months approached its close, and even prisoners of war and political culprits were released on parole if they wished to contest the laurel wreath of any championship, for to deprive them of the chance of winning such a distinction was thought a penalty too severe for a merely political offense. The ecstatic power of an Olympian triumph is well illustrated by the story of Diagoras, the Rhodian, who had been a famous champion in his younger days, and was present when his two sons won the entire *pentathlon*, i. e., carried off the five prizes for which the athletes of all Greece had been training during the four years preceding the sixty-first Olympiad. When the boys lifted their father up and carried him through the arena, the shouts of the assembled multitude were heard in the harbor of Patræ, at a distance of seven leagues, but Diagoras had heard nothing on earth after the herald's voice had proclaimed the names of the victors; "the gods," as Pindar says, "had granted that the happiest moment of his life should be his last." Would Diagoras have exchanged that moment for a week of those "beatific visions" which rewarded St. Dominic for his seven years' penance?

If any athlete received more than one prize of the same Olympiad, his victory was commemorated by a statue executed by the best contemporary sculptor of his native state. What a terrestrial Walhalla it must have been, that sacred mountain grove of Elis, where those statues were erected in the shade of majestic trees, while the summit of the hill and the open meadows were adorned by such masterpieces of Grecian architecture as the temple of Jupiter Olympius and the Pantheon of Callicrates! Besides the military drill-grounds and the public gymnasia, of which every hamlet had one or two, and where the complete apparatus for all possible sports was often combined with free baths and lecture-halls, the larger cities had associations for the promotion of the special favorite exercises, the brag-accomplishments of the rival towns. Wrestling, javelin-throwing, running, leaping, pitching the quoit, riding, driving, climbing ropes, shooting the arrow, were all practiced by as many amateur clubs, which commonly owned a race-course or a private hall.

Plato's Academia and Aristotle's Lyceum were both gymnastic institutions, where the patricians of Athens spent their leisure hours, and often joined in the exercises of the athletes. Our best citizens should emulate their example, and help to eradicate the lingering prejudice against the culture of the manly powers. A field-day, consecrated to Olympic games and the competitive gymnastics of the Turner-hall, should be the grandest yearly festival of a free nation.

In the mean time we must help our children the best way we can by giving them plenty of time for outdoor exercise, and providing them, according to our means, with some domestic substitutes for the gymnastic apparatus which, I trust, the next generation will find in every village hall and every town school.*

Children have a natural *penchant* for active exercises. Sloth is one of the vices we should drop from our catalogue of original sins. If a child were banished from the haunts of men, and left to grow up as a wild thing of the woods, he would turn out a self-made gymnast, though perhaps also in the original sense of the term, for gymnasium and gymnastics were derived from a word which means *naked*. Nature seems to deem the development of our limbs a matter of greater importance than their envelopment, and clothes are often, indeed, the first impediment to the free exercise of our motive organs. The regulation dress of the Swedish turners is, in this respect, also the best dress for children—a light jacket, wide trousers and shirts, and broad, low-heeled shoes; in-doors, and in summer-time, shoes and stockings should often be altogether dispensed with. Stephens, the celebrated English trainer, remarked that only men who have their toes perfectly straight will make first-rate runners and wrestlers, and this qualification is nowadays a privilege of country lads who are permitted (or obliged) to run around barefoot all summer. Considering the way we treat our feet, it must often puzzle us what our toes were made for, anyhow; but the antics of a baby in the cradle prove

---

* In 1825 Professor Beck opened in Northampton, Massachusetts, the first American school where gymnastics formed a branch of the regular curriculum. He has found followers, but, considering our progress in other directions, his wheat can not be said to have fallen on a fertile soil. Taking Massachusetts, Ohio, and North Carolina, as representative States of their respective sections, it seems that at present (1881) an average of three in every thousand North American schools pays any attention to physical education.

that the human foot is by nature semi-prehensile, and might be developed into a sort of under hand. Hindoo pickpockets "crib" with their toes, while they stand with folded arms in a crowd, and the Languedoc cork-gatherers ply their trade without a ladder, trusting their lives to the grasping power of their feet. The structural proportions of a new-born child also show a comparatively unimportant difference in the size of the lower and upper extremities; but, in the course of the first twelve years, this difference increases from $2:5$ to $1:3$, and often as much as $1:4$; in other words, while an infant's two arms weigh nearly as much as one of its legs, the arm-weight of a school-boy is often only one fourth of his leg-weight. The reason is that, of all the active exercise a child gets, nine tenths fall generally to the share of its lower extremities. A little child can not stand erect; the task of supporting the weight of the whole body on two feet exceeds its untried strength. But in local progression we do more: taking a step means to support and propel, or even lift, the whole body by means of the foot remaining on the ground. In running up and down stairs, to school and back, and here and there about the house, the legs of the laziest school-boy perform that feat about eight thousand times a day. What have his arms done in the mean while? Carried a chair across the room, perhaps, or elevated so and so many spoonfuls of hash from the plate to a place six inches farther up, besides supporting the weight of three or four ounces of clothing. To equalize this difference should therefore be the primary object of physical culture, for the harmonious structure of all its parts is an essential condition of a perfectly developed body. No malformation is more common in city recruits than a narrow chest. Besides spear-throwing, of which I shall speak further on, any exercise promoting the development of the shoulder-muscles will tend to expand the chest, and thus remove the chief predisposing cause of consumption. In a climate where the first four years of a child's life have to be passed mostly indoors, a special room of a spacious house or a corner reservation of a small nursery should be set apart for arm-exercises—hurling, swinging, and lifting. The arrangements for the propulsive part of the good work need not go beyond an old bolster and a cushion-target, but the *grapple-swing* should be both safe and handy— a pair of swinging-rings suspended at a height of about four feet from the floor above a stratum of old quilts and carpets. In London, and in some of our Northeastern cities, *health-lifts* for children can now be got very cheap; weighted buckets, however, or sand-bags with strap-handles, will serve nearly the same purpose; and smaller bags of that kind may be used for various dumb-bell exercises. A plurality of young gymnasts can vary the programme by throwing such bags to each other and catching them with outstretched arms. In a suitable locality I would add a knotted rope, fastened to the ceiling by means of a screw-hook, and hanging down in a single or double chain, which children soon learn to climb by the hand-over-hand process, thus strengthening the triceps and flexor muscles, to whose development the quadrumana owe their peculiar arm-power. A full-grown man who has passed his life behind the counter will find it rather difficult to raise his body by the contraction of his arm-muscles, but, unless Darwin is right, Heaven must have intended us to pursue the culture of our higher virtues in the tree-tops, after the manner of the gymnosophists, for a young child acquires all climbing tricks with a quite amazing facility—much readier, in fact, than the art of biped progression, whose chief difficulty consists, perhaps, in the necessity of preserving the equilibrium. The knots should be far enough apart to tempt an enterprising climber to dispense with their use now and then, and rely on the power of his grasp by seizing the rope at the interspaces; and this exercise should be especially encouraged, for the strength and suppleness of the wrist-joint will considerably facilitate the attainment of "polytechnic skill," as modern Jacks-of-all-trades begin to call their versatile handiness. Nay, the Rev. Salzmann holds that the ancient practice of hand-shaking was originally suggested by the wish to ascertain the wrist-power and consequent wrestling capacity of a stranger. As to the rest, negative precautions will generally suffice for the first three or four years. Diminish the danger of a fall by padding the floor of your nursery-gymnasium, and the restless mobility of your pupils will generally save you the trouble of initiating them in the rudiments of hopping and tumbling. But make it a rule with all hired or amateur nursery-maids that the children must not be carried more than is absolutely necessary.

In long winters it can do no harm, now and then, to let the youngsters turn the hall into a race-course; but, with the first warm weather, the arena should be removed to the next playground—a garden-lane, or a vacant lot without rubbish-heaps, if the Park Commissioners are too proscriptive. In its general invigorating effect on the organic system, running surpasses every other kind of exercise. Among the contests of the palæstra it ranked above wrestling and boxing; for more than two hundred years the Olympic games consisted, indeed, exclusively of foot-races, and the chronological era of Greece dated from the year when the

Elean Corœbus defeated his Peloponnesian competitors in the long-distance match. The swift-footedness of Achilles is mentioned as often as his name occurs in the "Iliad"; and, according to the Scandinavian Saga, the champions of Jötunheim distanced even the henchman of Thor in a foot-race. Next to a smooth and perfectly level lawn, a firm beach is the best race-course, and, after a warm day, it is a luxury to the martyred feet of a city boy to tread the cool sand with his naked soles. Fast running is, on the whole, a more valuable accomplishment than long walking, for no one knows when he may owe his life, and more than his life, to the ability of outrunning a pursuer or a fugitive scoundrel; but walking and trotting matches against time will help to cure our children of that miserable snail-pace which has come to be the fashion of every public promenade. Reduced to a funeral-march, the "regulation walk" loses half its value—the hygienic value of the only kind of out-door exercise which the children of the upper ten or twenty can count upon. Who could wish a prettier sight than a bevy of school-girls, flitting by with fluttering flounces, like dancers keeping step to a merry tune? If mothers knew all the charms of *animated* beauty, they would not think it "more becoming" to turn their children into tortoises. Nor would they fear that they would "run themselves into a consumption," if they knew what real running means, and what the motive organs of a human being are capable of. Mexico has ceased to be a *terra incognita* to Yankee tourists, and most visitors to the upland cities will remember the army of hucksters and poulterers who every forenoon turn the main plaza into an agricultural fair. If you will take a morning walk on one of the sand-roads that diverge from the south gate of

Puebla, you may see those hucksters coming in *at a trot*, girls in their teens many of them, and loaded with sacks and baskets; and upon inquiry you will learn that most of them come from the valley of Tehuacan, from a distance of ten or twelve English miles. The *zagal*, or post-boy of a Spanish mail-coach, carries nothing but a light whip, but he has not only to keep pace with a team of galloping horses for hour after hour, but has to run zigzag, adjusting a strap here, picking up a handkerchief there, and frequently entertains the travelers with a series of hand-springs, in order to earn an extra *medio* or two—not to mention the Grecian *hemerodromes*, who could distance a horse on the long run, and had often to cross rivers and lakes on their bee-line routes.

An excellent system of training was that of the old Turkish Jenidji-begs, or drill-masters of the Janizary cadets, who made young boys practice lance-throwing with a spear that *exceeded* the common javelin both in size and weight—" because, after they had become proficient in the use of such a heavy implement, the army-spear would be a mere feather in their hands." On the same principle the knee-muscles may be strengthened by a simple manœuvre without the use of any apparatus. Bend the left leg in a right angle, extending the right leg horizontally, and lower the body till your right heel nearly touches the ground. Now rise by straightening the left leg, with the right still extended horizontally, and without letting your hands or your right heel touch the ground. Then squat down as before, extend the left leg this time and rise on the right, and so on until the weight of the body has been raised twenty or thirty times by the effort of either knee-joint without the aid of the other. A moderate proficiency in this exercise will enable girls and city boys to walk up-hill for hours with the ease of a Tyrolese goat-herd.

In classifying gymnastics after the degree of their usefulness, a prominent place should be assigned to leaping, especially high leaping, an exercise which imparts a powerful stimulus to the digestive organs, and, combined with the shock of the descent, exerts an invigorating influence on the nervous system in general. The leaping-gauge of the Turner-hall consists of two upright posts with pegs and a cord stretched from post to post. Every peg is marked with a figure indicating the number of inches from the ground, and by raising or lowering the cord each gymnast can measure his jumping capacity and keep tally of his score in a certain number of leaps. Competition imparts to this sport an incentive which may be put to as good account in gymnastics as in mental exercises, and is certainly preferable to the only other method of stimulating the zeal of young pupils. Personal ambition, according to the ethics of a certain class of pedagogues, is inconsistent with the spirit of true Christian humility, and should be quelled rather than fomented; in dealing with unruly youngsters they have consequently to resort to the only alternative, slavish fear, enforced by punishments and espionage. For the nonce, that system answers its purpose quite as well as the emulation-method; as to future results, your choice must depend upon the main question of modern education, Are we to form men, or canting sneaks?

A quadruped has an evident advantage over a biped jumper, but practice will do wonders. Leonardo da Vinci often astounded his visitors by jumping to the ceiling and knocking his feet against the bells of a glass chandelier, and a private soldier of Vandamme's cuirassiers even leaped over the tutelar deity of a brass fountain on the Frankfort market-square. But the champion jumper of modern times was Joe Ireland, a native of Beverley in Yorkshire. In his eighteenth year, "without any assistance, trick, or deception," he leaped over nine horses standing side by side and a man seated on the middle horse. He could clear a string held fourteen feet high, and once kicked a bladder hanging sixteen feet from the ground.* In horizontal leaps our turners can not beat the record of antiquity: a Spartan once cleared fifty-two feet, and a native of Crotona even fifty five. Nor would any modern filibusters be likely to emulate the trick of the Teuton freebooters who crossed the Alps during the consulate of Caius Marius: Finding the Roman battle-front inexpugnable, they attempted to force the fight by vaulting with the aid of their *framæ* or leaping-poles over a triple row of mail-clad spearmen.

Hurling is the gymnastic specific for pulmonary complaints; and the best possible exercise for so many hectic and narrow-chested boys of our larger cities would be the game of *Ger-werfen*, as the turners call it—spear-throwing at a fixed or movable mark. It is a most diverting sport after a week's practice has hardened the flexor muscles against the shock of propelling the larger spears. The missile is a lance of some tough wood (ash and hickory preferred), about ten feet long and one and a half inch in diameter, terminating in a blunt iron knob to steady the throw and keep the wood from splintering. A heavy post with a movable top-piece (the "Ger-block") forms the target, the head-shaped top being secured by means of a stout cramp-hinge that permits it to turn over, but not to fall down

---

* Strutt's "Plays and Pastimes," p. 176.

—distance, all the way from ten to forty paces. Grasp the spear near the middle, raise it to the height of your ear, plant the left foot firmly on the ground, the right knee slightly bent, fix your eye on the target, lean back and let drive. If you hit the log squarely in the center or a trifle higher up, it will topple over, but, still hanging by the cramp-hinge, can be quickly adjusted for the next thrower. A feeble hit will not stir the ponderous Ger-block; the spear has to impinge with the force of a sixty-pound blow, so that a successful throw is also an athletic triumph. The German Ger-throwers are generally lads after the heart of Charles Reade—ambidexterous boys, whose either-handed strength and skill illustrate the fact that the antiquity of a prejudice proves nothing in its favor. As the least vacillation in the act of throwing would derange the aim, this exercise imparts a perfect command over the balance of the body, besides improving the faculty of measuring distances by the eye. It is, indeed, surprising how soon gymnastics of this sort will impart an easy deportment and graceful manners even to boys in their lubber-years—"*Nur aus vollendeter Kraft strahlet die Anmuth hervor*," as Goethe explains it: "The highest grace is the outcome of consummate strength."

Climbing, too, calls into action nearly every muscle of the human body, and should be encouraged, though at the expense of a pair of summer pants or summer birds, as the possibility of accidents is more than outweighed by the sure gain in physical self-reliance. There is a deep truth in the apparent paradox that it is the best plan *not* to avoid dangers and difficulties that can be mastered. In the voluntary risks of the gymnasium the athlete pays an insurance policy against future dangers. In a man's life there will always come moments when the woe and weal of years depend on firm nerves and a strong hand, and such moments prove the value of a system of training which teaches children to treat danger as a mechanical problem. The operation of the same cause may be traced in the *realistic* influence which the culture of the manly powers generally exerts on the human mind. Having learned to rely on their personal strength and judgment under circumstances where shams are peculiarly unavailing, gymnasts will generally be men of self-help; practical, rather apt to believe in the competence of human reason and human virtue and to question the utility of a pious fraud.

On rainy days an in-door gymnasium is as useful as a private library. Where wood is cheap, the aggregate cost of the following apparatus need not exceed fifty dollars: 1. A spring-board and leaping-gauge; 2. An inclined ladder; 3. An horizontal bar; 4. Swinging-rings; 5. A vaulting-horse (rough hewed); 6. A chest-expander (elastic band with handles); and, 7. A pair of Indian clubs. Buckets filled with shot or pig-iron will do for a health-lift. With this simple apparatus an infinite variety of health-giving exercises may be performed without much risk; on the horizontal bar alone Jahn and Salzmann enumerate not less than one hundred and twenty different movements, most of which have proved very useful in correcting special malformations. For general hygienic purposes a much smaller number will be sufficient, especially where the neighborhood affords an opportunity for occasional out-door sports; for an in-door gymnasium is, after all, only a preparatory school, or at best a substitute for the palæstra of Nature—the woods, the sea-shore, and the cliffs of a rocky mountain-range. But in large cities even the poorest ought to procure a few gymnastic implements; no dyspeptic should be without a spring-board and some sort of health-lift.

The victims of asthma would throw a considerable quantity of physic to the dogs if they knew the value of a mechanical specific—a few minutes' exercise with the *balance-stick*, an apparatus which any man can manufacture in half an hour, and at an expense representing the value of an old broomstick and a yard of copper wire. Take a straight stick, about six feet long and one inch in diameter, and mark it from end to end with deep notches at regular intervals, say two inches apart, with smaller subdivisions, as on the beam of a lever-balance. Then get a ten-pound lump of pig-iron, or a large stone, and gird it with a piece of stout wire, so as to let one end of the wire project in the form of a hook. The exercise consists in grasping the stick at one end, stretching out arm and stick horizontally like a rapier at a home-thrust; then draw your arm back, still keeping the stick rigidly horizontal, make your hand touch your chin, thrust it out again, draw back, and so on, till the fore-arm moves rapidly on a steady fulcrum. Next *load* the stick—i. e., hook the stone to one of the notches; every inch farther out will increase the weight by several pounds. Hook it to one of the middle notches, and try to move your arm as before. It will be hard work now to keep the stick horizontal; even a strong man will find that the effort reacts powerfully on his lungs: he will puff as if the respiratory engine were working under high pressure. On the same principle, the lungs of a half-drowned man may be set awork by moving the arms up and down like pump-handles. But the weighted stick, bearing against the sinews of the fore-arm, still increases this effect, and overcomes the stricture of the asthmatic spasm, as the movement of the loose arms relieves the torpor of the drowning-asphyxia. With the aid of this mechanical *palliative* (for death is the only radical asthma-cure) the distress of the spasm can be relieved before the actual dyspnœa or breathlessness has begun, and, after ten or twelve resolute efforts, the feeling of oppression will generally subside and the lungs resume their work as if nothing had happened. Daily exercise with the balance-stick is sure to diminish the frequency of the attacks, and, if begun in time, would probably cure children from an hereditary tendency of this sort. Two years ago I sent this receipt to an asthma-martyr whom the narcotic-vapor cure did not save from a weekly repetition of all the horrors of strangulation. He has now lengthened the period of his complaint from a week to an average of forty days, and assured me that even a few minutes' exercise with a six-pound weight has saved him many a sleepless night.

Lifting and carrying weights was a favorite exercise with the ancient athletes, and our modern rustics are still very apt to estimate a man's strength by his lifting capacity. The "best man" of a Yorkshire parish is generally he who can shoulder the heaviest bag and carry it farthest and with the firmest step. Feats of this sort require certainly a sound constitution in every way; weak lungs, especially, are sure to tell, but the main strain bears upon the thighs and the small of the back: a good lifter has to be a strong-boned man, and will generally make a good wrestler and rider. Weak-backed children will, therefore, derive much benefit from the various exercises with hand-weights and lifting-straps, and, indeed, from any labor involving the addition of an extra burden to the natural weight of the body. Heavy lifts require some precaution against strains—a waist-belt, and unflinching steadiness in rising from a stooping position; but it should be remembered that *rupture* (hernia)—generally ascribed to the effects of over-lifts—results more frequently from the shock of a fall, and a predisposing defect of the abdominal teguments. The history of the lifting-cure records not a single instance of a rupture having *originated* from the often enormous feats of professional gymnasts, or the more dangerous efforts of enthusiastic beginners. As a general rule, it may be relied upon that a perfectly sound child can not over-lift himself before his strength gives way—I mean, before the yielding of his muscles and sinews simply compels him to drop the burden. Here, too, the achievements of ancient and modern Samsons illustrate the tenacity of the human frame and its marvelous capacity for development. The credibility of the Gaza story depends somewhat upon the size of those city gates; but there is no doubt that Thomas Topham, of Surrey, once shouldered a sentry-box containing a stove, a bench, and a sleeping watchman, and carried his burden to a suburban cemetery. Dr. Winship, of Boston, lifted twenty-nine hundred pounds with the aid of shoulder-straps; and, unless the historians of Magna Græcia were afflicted with an abnormal development of the myth-making faculty, it would seem that their countryman Milo carried a bull-calf around the arena, and thus carried it every day till he could tote a full-grown steer. If the story is even half true, we need not wonder that Milo's powers as a wrestler put a temporary stop to that sport as a branch of the Olympian games, since "no man or god durst accept his challenge."

Wrestling is still the chief accomplishment of the

Swiss village champions, and would be the favorite pastime of our rural districts if it had not been kept down by our sickly prejudice against all rough-and-ready sports. Fifteen centuries ago the Olympic games were abolished by the decree of a Christian emperor; the moralists of Old England have tabooed pugilism; our sabbatarians now include even wrestling among the "blackguard sports"; and Frederick Gerstaecker predicts that the American Inquisition of a future century will suppress skating and ball-playing "as giving an undue ascendency to the animal energies over the moral part of our nature." For such a century's sake we should hope that the Patagonian savages will prove unconquerable, for a year's *life* among healthy beasts would be a blessed relief from a long sojourn in the land of an unmanned nation.

But I trust that the propaganda of the Turnbund will save us from such a fate. What a stimulus it would give to manly sports and manly virtues, nay, to the physical regeneration of the human race, if we could make their yearly assembly a national festival! The river-meadows of Chattanooga, or the mountain amphitheatre near Huntsville, Alabama, would make a first-class Olympia, and our Indian summer would be a ready-made "weather-truce," without an expensive burnt-offering to the sun. Olives, it is true, do not flourish on our soil; our mercenary souls need other inducements; but the rent of reserved seats and camp-tents would enable us to gild the crowns of the several victors. Imagine the athletes of every village training for those prizes—thousands of boy-topers turning gymnasts, ward delegates running for something besides office, and the members of a Young Men's Association seeking paradise on this side of the grave!

With the decadence of athletic sports, games of skill come generally into favor; hence, perhaps, the revival of archery in the United States, and the pandemic spread of certain amusements which are properly ladies' plays. Riding has gone almost out of fashion, though few sportsmen will gainsay me if I assert that a day in the saddle is worth a week of other *sedentary* pursuits. A Mexican boy would part as soon with an arm as with his horse, and I never saw a finer picture of exultant health than a cavalcade of *muchachos* dashing out into the prairie at full speed, whooping and cheering, though perhaps on their way to school or to a *funcion* of some national saint. The deportment of such little equestrians is distinguished by a certain chivalrous frankness, and the word *chivalry* itself, as well as the German *Ritter* ("caballero"), was originally derived from horse-riding. The rider's mangement of his nag may tend to develop the domineering, the princely traits of human nature, though probably at the expense of a humbler virtue or two; in Spanish America, at least, the experience of Indian agents and Indian school-teachers has shown that the pedestrian red-skins are generally more manageable than their mounted *compadres*.

The lovers of aquatic sports may combine a useful accomplishment with the best relief from the midsummer martyrdom of our large cities. The art of swimming adds as much to the pleasure of bathing as it does to its healthfulness; but it has often puzzled me that with the human animal that should be an art which is a natural faculty of *all* other mammals. Dr. Andersson's theory is probably the right solution of the riddle. He noticed that to the young negroes of Sierra Leone swimming comes almost as natural as walking (in which attainment they are also rather precocious), and he concludes that the disability of a white man's child arises chiefly from a general want of vigor. Our mobile arms and paddle-like hands are better swimming implements than the drumstick legs of a dog; but our muscular debility more than counteracts these advantages. The limbs of a child are swathed, confined in tight clothes, kept year after year in compulsory inactivity, till, in proportion to its size, the nursling of civilization is the weakliest of living creatures. After exercise has developed the defective muscles, a swimmer can hardly understand how he could ever be in dread of deep water, swimming seems so easy: the faculty of floating, as it appears to him, is an inalienable attribute of a human creature, requiring neither art nor anything like a great effort except in swimming against the stream; he would undertake to study, read, or dream in a calm sea, and let the body take care of itself. The Marquesas-Islanders witnessed the struggles of a sinking English sailor with mute astonishment, and neglected to help him, utterly incapable of realizing the fact that a full-grown man could be in danger of drowning.

In sixteen provinces of the Roman Empire every larger town had a free bath or two, and our entire neglect of this branch of public hygiene is certainly the ugliest feature of our boasted civilization; but our children at least might make shift with the natural bathing facilities which can be reached by a short excursion beyond the precincts of all but the unluckiest cities. A cool bath at the end of a sweltering day can be delightful enough to reconcile a poor city slave to his misery; the sensation of floating along with the rhythm of a dancing current admits no comparison with any *terra firma* pleasure, and awakens instincts of the human soul which may date from the life of our marine ancestors in the days of the Devonian fore-world. But such enjoyments are the privilege of the aquatic gymnast, and no swimmer should deem it below his dignity to imitate the example of the elder Cato, who taught his sons to dive and traverse rapid rivers. I know that a swimming-school is not always a favorite resort of a young child; weakly youngsters are apt to prefer a sponge-bath; but I agree with the Baptists, that immersion alone will save us. The way of the beginner is hard, but the reward is worth the price. No boy who has learned to "tread water" or to "take a header" from a high bank would exchange the wild joy of his sport for all the taffy of a tame Sunday-school picnic. And it is a great mistake to suppose that hardy habits would harden the character; on the contrary, the bravest lad of a parish can generally be known by his cheerfulness and his frank good-nature, and in after-years will be apt to meet the billows of life with a joyous zeal rather than with a shivering "resignation." I am often tempted to quote the remark of a French training-ship surgeon, of blunt speech, but with a sharp eye for the character-traits of his young countrymen: "If I had my own way," said he, "every boy in the marine should serve an apprenticeship in the rigging, and learn to rough it before he gets a soft berth. The lads that have grown up before the mast make the best men in every sense of the word, brave, honest fellows most of them; while the cabin-boys, who have been pampered with tidbits and soft jobs generally, turn out" (I won't risk a literal translation) "prevaricating puppies," or words to that effect.

Per aspera ad astra, and a very important branch of gymnastic education might be included under the head of hard work or voluntary labor. Labor with a practical purpose is not only more visibly useful but more agreeable than mere crank-work at the horizontal bar, and it is sometimes advisable to beguile ourselves into a strenuous and long-continued physical effort. For what we call vice or evil propensities is often nothing but misdirected energy, vital force exploding in the wrong direction for want of a better outlet. The sensible remedy is not to anathematize such energies, but to let our muscular system absorb them by engaging in some entertaining out-door business requiring a good deal of heavy work. In summer-time there will be no lack of such jobs: interest your *enfant terrible* in horticulture; make him transplant shade-trees and dig ditches; send him to the gravel-pit, and let him fill his wheelbarrow with sand and his pockets with geological specimens. Or enlist his constructiveness: set him to build a garden-wall, and quarry his own building material in the next ravine. During the progress of the good work the hours will vanish magically, and so will the evil propensities. Novel-reading girls can generally be cured with a butterfly-catcher; entomology and sentimentalism are not concomitant manias.

It has often been observed as a curious phenomenon that the vilest young hoodlums are found in the middle-sized towns. I believe I could suggest an explanation: In very large cities, as well as in the woods and mountains, they find something else to do. A New York street Arab is often addicted to sharp practice, but not often to degrading vices. He can't afford to be vicious: sensuality weakens; physical vigor is a stock-in-trade; the fierceness of competition compels him to use every advantage. For the same reason a training oarsman is generally an exemplar of all manly virtues; to him experience has demonstrated the *temporal* disadvantages of vice, an argument whose cogency somehow conquers objections that resist the most eloquent *argumenta ad fidem*. Moreover, such virtues with a business purpose are liable to become habits. If we could keep a record of the longevity of our university crews, we would probably find that the victors outlive the often vanquished; the champions of Olympia (with the exception of the cestus-fighters) generally attained to a good old age.

It is, indeed, a pity that oar-contests should be confined to our lake-shore cities and a few college towns; as an athletic exercise rowing is out and out superior to ball-playing and skating, and a sovereign remedy for many disorders of the respiratory organs. Venice has all the topographical characteristics of a consumption town—stagnant lagoons, damp buildings, dark and narrow streets—and yet the lower classes of her population are remarkably free from pulmonary affections—they have a gondolier in nearly ever family. The watermen of the Thames, too, are generally long-lived, in spite of being so much exposed to wet and cold. If I had to limit a child to two kinds of out-door exercises, I would choose running and rowing: the one does for the legs and the stomach what the other does for the arms and the lungs.

It is said that Cyrus advised his countrymen "never to eat but after labor," and, as a general rule, the best time for out-door exercise is certainly rather before than after meals; but gymnastics of the heroic kind may induce a degree of fatigue which decreases the appetite instead of stimulating it, and in summer it is by far the best plan to take the last meal in the afternoon, and postpone athletic sports to the cooler hours of the evening. In moonlit nights, out-door games may be continued for several hours after sunset. A nearly infallible receipt for pleasant dreams is a light supper, followed by competitive gymnastics in the presence of (somebody's) sisters and cousins. In stress of circumstances, though, the fair witnesses can be dispensed with. Even an in-door gymnasium will answer the main purpose; it is the relaxation of the strained sinews which makes rest sweet; the soul seems to revel in a conscious sense of health to come. It is a fact that a man may be "too tired to sleep"; but that sort of insomnia is always a sign of general debility. Our latter-day sports are not likely to hurt a healthy boy through excess of exercise. We hear of people having "killed themselves with hard work"; but, if their habits were otherwise correct and their diet not altogether insufficient, they must have worked hard indeed, and with *suicidal intent,* I am tempted to say, as we have no single word for *Lebensmüde*—the reckless contempt of life which can make men deaf to the voice of their physical conscience. The Manitoba lumbermen ply their hard trade cheerfully for ten hours a day for months together, and the pastoral nomads of the Caspian steppes often keep their boys in the saddle for two days and two nights.

It can do no harm to let girls join in the athletic sports of their brothers; though in their case an harmonious structural development is of more importance than the attainment of muscular strength. Their natural vocation exempts them from the necessity of engaging in violent exercises, and the experience of every nation has confirmed the somewhat obscure biological fact that a child's bodily constitution depends chiefly on that of his paternal relatives. A weakling can never become the father of robust children; while a delicate but otherwise healthy woman may give birth to an infant Hercules. But, for boys, the most thorough physical education is the best; a child can never be too weakly to profit by gymnastic exercises. If the culture of the bodily faculties were made a regular branch of public education, robust strength would be the rule and debility the rare exception. The puniness and sickliness of the vast plurality of our city boys are indeed something altogether abnormal. If our primogenitor (as we have no reason to doubt) surpassed the other primates of the animal kingdom in strength as much as he still exceeds them in size, he must have been fully able to hold his own against any beast of prey. Dr. Clarke Abel's undoubtedly authentic description of an orang-outang hunt near Rangoon, on the northwest coast of Sumatra, reads like an episode from the "Lay of the Nibelungen," rather than like the account of a conscientious and scientific observer. With five bullets in his body, the hairy half-man still leaped from tree to tree with the agility of a panther, survived the fall of the last tree, and, though crippled by a shower of blows, snatched a spear from the hands of his chief assailant and broke it like a rotten stick. On his campaign against a horde of northern barbarians, one of Trajan's generals attempted to scare, or at least to astonish, the natives by shipping a troop of lions across the Danube. But the children of Nature declined to marvel: "They mistook them for dogs," says the historian, "and knocked their brains out." Even after the middle of the fourteenth century the levy of a small German burgh could turn out more athletes than the combined armies of the present empire; the Margrave of Nuremberg could at any time muster ten thousand men, every one of whom was able to wear and use accoutrements that would crush a so-called strong man of the present day. In the armories of Vienna, Brunswick, and Strasburg there are coats of mail which a modern porter would hesitate to shoulder without the assistance of a comrade.

And yet these mediæval Samsons were the exclusive product of the drill-ground; physical vigor was not valued as the foundation of health and happiness, but rather as a means of military efficiency; the guardians of public education merely connived at such things; and, when the invention of gunpowder diminished the importance of personal prowess, our anti-natural dogmas accomplished their tendency in the rapid physical corruption of their devotees. The dull and gloomy slavery of the monasteries was transferred to the management of all educational institutions; for several centuries the bodily rights of the poor convent-pupils were not only disregarded but willfully depreciated. Educational influences became the chief cause of physical degeneracy, and the superficialness of our reformatory measures proves that we have not yet recognized the root of the evil.

But the voice of Nature has repeated its protest in the yearnings of every new generation. Our children still long for out-door life, for active exercise, for the free development of every bodily faculty; and, if we cease to suppress those instincts, the regenerative tendency of Nature will soon assert itself, and the time may come when man will be once more the physical as well as mental superior of his fellow-creatures.

CHAPTER V.

CLOTHING.

"No better traveling habit than *hardy habits.*"—Sir Samuel Baker.

The capacity of our ancestors to accommodate themselves to every climate depended not only on their physiological faculty of adaptation, but also on their skill in protecting themselves by artificial means from the inclemency of the higher latitudes. Houses and clothes are a blessing if they answer this purpose by a close imitation of Nature's own plan in sheltering her children from atmospheric vicissitudes; but in degree as they deviate from that plan their hygienic disadvantages balance, or even outweigh, the gain in other respects. A swallow's nest protects her brood from cold and rain without debarring them from the fresh air; a human domicile, too, should combine comfort with the advantage of perfect ventilation; and our clothes, like the fur of a squirrel or the feather-mantle of a hawk, should keep us warm and dry without interfering with the cutaneous excretions and the free movement of our limbs.

Measured by these standards, the winter dress of an American school-boy is nearly the best, the summer dress of the average American, French, and German nursling about the worst, that could possibly be devised. At an age when the rapid development of the whole organism requires the utmost freedom of movement, our children are kept in the fetters of garments that check the activity of the body in every way: swaddling-clothes, undershirts, overshirts, neck-wrappers, trailing gowns, garnitures, flounces, and shawls reduce the helpless *homunculus* to a bundle of dry goods, unable to move or turn, incapable of relieving or intimating its uneasiness in any way save by the use of its squealing apparatus, and consequently squealing violently from morning till night. Out-doors, in the baby-carriage, "cold draughts" have to be guarded against, and a load of extra wrappers completely counteract the benefit of the fresh air; faint with nausea and suffocating heat the little mummy lies motionless on its back, resplendent in its white surplice, a fit candidate for the honors of a life whose every movement of a natural impulse will be suppressed as a revival of barbarism and an insurrection against the statutes of an orthodox community. Hence, in a great degree, the disproportionate mortality, in all northern countries of Christendom, among infants under two years. In Spanish America, where infantile diseases are as rare as in Hindostan, babies of all classes and all sizes toddle about *naked*, nearly the year round; and the Indians of Tamaulipas, between Tampico and Matamoras, raise an astonishing number of brown bantlings who are never troubled with clothes till they are big enough to carry garden-stuff to a city where the police enforces the apron regulation.

But Mrs. Grundy—a person's pinafore—and the carpet? Well, get a lot of short linen hose, rather loose about the hips and tied around the waist or buttoned to the skirts of a short frock. Change them as often as you like. Wholesale they could be made for a dollar and washed for a quarter a dozen. Out-doors add a pair of stockings with canvas soles, and perhaps little rubber boots on wet days, but no cap or shawl before October, and under no circumstances any swaddles or baby night-gowns. Let us get rid of the "draught" superstition; catarrhs are not taken by any creature of the open air, not by the fisherman's boy, paddling around in the surf and sitting barefooted in a wet canoe or bareheaded on the windward cliffs, but by the cachectic cadets of the tenement-barracks, where the same air is breathed and rebreathed by the diseased lungs of a regiment of voluntary prisoners.*

After the first frost, a cap with ear-flaps, double stockings, and mittens out-doors can do no harm. A warm shirt and two quilt-blankets will be enough in all but the coldest nights, and (if I had not seen the thing done I should commit an outrage on common sense by thinking it necessary to mention it) the face of a sleeping child should never be covered with a shawl, nor—when flies are very troublesome—with anything thicker than the lightest gauze handkerchief. "A great store of clothes," says Lord Bacon, "either upon the bed or the back, relaxes the body"; and every observant parent must have noticed that school-children complain a hundred times of being overdressed for once that they ask for additional or warmer clothing. Indeed, only dire habit can reconcile us to the mass of trappings and wrappings which fashion and effeminacy load us with. Five hundred millions of our fellow-men wear scarcely any clothing—not in Africa

---

* "I shall not attempt to explain why 'damp clothes' occasion colds rather than wet ones, because I doubt the fact. I imagine that neither the one nor the other contributes to this effect, and that the causes of colds are totally independent of wet and even of cold."—(Ben Franklin's "Essays," p. 216.)

and Southern Asia only, but in cold Patagonia and the by no means genial latitudes of the Norfolk Islands. The mantle of the Roman peasant was laid aside in cold weather and generally at the beginning of the day's work. The sculptures of Rome and Greece abound with the representations of nude hunters, shepherds, and artisans. On the friezes of Pompey and the countless vases and entablatures of the Museo Borbonico and the Vatican collection, children, almost without any exception, appear *in naturalibus*. The very word *gymnasium* was derived from γυμνόσ, *naked;* and there is every reason to believe that the *toga virilis*, like the *toga prætexta*, was worn only on state occasions. Henry's "History of Great Britain" (vol. i, pp. 468, 469) leaves hardly any doubt that the ancient Britons, Picts, and Scots were either wholly or almost naked, "unless their custom of painting their bodies can be considered as clothing." Nor did the south Britons and Romans go naked from poverty, like Darwin's Firelanders. They had clothes, but they reserved them for emergencies, and, though our advanced notions of decency and cleanliness might not permit us to emulate their example, I suspect that, from May to November, the lightest suit of clothes is, from an hygienic standpoint, about the best. The body breathes through the pores as well as through the lungs, and heavy garments obstruct the cutaneous exhalations quite as much as the atmosphere of an over-heated room impedes the process of respiration, and it has been found by actual experiments that the weight of a mantle or heavy coat with woolen shirts and other underwear diminishes the respiratory capacity of the lungs from twenty to twenty-five per cent. (Coale's "Hints on Health," p. 104.)

Besides, it seems that fresh air exercises on the

human skin a certain tonic influence, of which the wearer of thick woolen garments deprives his body. Benjamin Franklin proposed to prevent colds, and even small-pox, by *air-baths*, and found that he could relieve insomnia by simply removing the bedclothes for a couple of minutes. "I rise early almost every morning," says he, "and sit in my chamber without any clothes whatever, half an hour or an hour, according to the season, either reading or writing. This practice is not the least painful but, on the contrary, agreeable, and if I return to bed afterward, before I dress myself, as it sometimes happens, I make a supplement to my night's rest of one or two hours of the most pleasing sleep that can be imagined. ("A New Mode of Bathing," Franklin's "Essays," p. 215.)

Nor should we forget the incidental advantages of hardy habits, their invigorating influence on the constitution in general and on the digestive system in particular, nor the fact that effeminacy defeats its own object and exposes its slaves to sufferings unknown to the sons of the wilderness. He who restricts himself to a minimum of clothes in summer-time will find an extra shirt or a plaid and a pair of mittens a sufficient protection from almost any weather. The Indians of the Tehuantepec highlands, who work the year round in a breech-clout and a palmetto hat, ascend the icy summit regions of the Sierra Madre with a threadbare blanket as their only cover from cold winds and night frosts; and our own red-skins prefer an old buffalo-robe to the best tight-fitting garments, and invariably tear the seams of the store-clothes they buy at the post-agencies—to make them "lighter," ventilate them, as it were. Nay, the post-trader of Fort Richardson, on the upper Brazos, assured me that his Kiowa customers never bought a suit of clothes without cutting the seat out of the pantaloons and slitting the coats from the armpits down to the skirts!

If an out-door laborer leaves a warm house on a cold morning, the first contact with the open air is anything but agreeable, but after half an hour's exercise the body warms up from within, and this animal caloric can make a heavy suit of clothes as oppressive in winter as in midsummer; the gaseous excretions of the skin, after saturating the confined air, are condensed and thus effectually checked—the body has to forego the benefits of cutaneous respiration. And herein consists the difference between our artificial fleece and the hairy coat of a wild beast: fur and wool retain the animal warmth but emit the cutaneous vapors; a close woven coat stops both. The process of tanning, too, stops the pores of the fur-skin, and I have often wondered why our dress-reformers have never tried to construct a fur coat on the brush-maker's plan—fastening the hair in little bunches on some strong, net-like texture. By spreading outward, the hair would present the even surface of the natural fur, and make such a porous brush coat nearly as warm as a common pelisse. Thus far the same end has been most nearly attained by the triple blouse of the Havre 'longshoremen—three linen jackets; the first and third as smooth as a shirt, but the middle one *ruffled*, i. e., gathered up in a series of open plaits like a mediæval lace collar. This arrangement prevents a "tight fit," and leaves a considerable space on both sides of the middle blouse, and, air being a bad conductor, the three blouses, weighing about three pounds apiece, are actually warmer than a twelve-pound overcoat of thick broadcloth, but fitting the back like the cover of a pin-cushion. On going to work, the *porte-faix* removes one or two of his blouses, according to the state of the weather, as the American school-boy takes off his comforter and unbuttons his jacket before going in for a snow-ball fight.

A jacket or a short blouse is out and out more sensible than our cumbersome overcoats or the unspeakable tangle-work of frippery and flounces, cross- and-lengthwise wrappings, and intricate fastenings that still form the winter dress of a fashionable lady. The women of Scandinavia and New England (Jenny Lind, Mrs. Everett, Dr. Mary Safford-Blake, etc.) can claim the honor of having initiated the opposition movement that bids fair to abate the grievance in the course of another generation or two, having already exploded the chief outrages on hygienic and artistic common sense— corsets and the crinoline. Mrs. Abba G. Woolson's "Dress Reform" should be the sartorial text-book of every girl's mother.

The Turks and Hollanders, though differing so widely in their general mode of life, agree in preferring warm clothes to heated rooms, and, when the in-door atmosphere can be made tolerable only by air-tight window-sashes and glowing stoves, it is a curious question whether a warmer dress would not, on the whole, be the lesser evil. It would save fuel, sick-headaches, and constipation, and by adding or removing an extra blouse, *à la Normandie*, the several occupants of a moderately warmed room might exactly adapt the temperature to their individual feelings. A German author, who admits hardly any excuse for excluding the fresh air from a sitting-room, proposes an ingenious remedy for *cold hands*—the only cogent objection to an open study-window: a box writing-desk, namely, with a double lid, the writing-board resting on top of a box full of hot sand, that can be warmed in a common baking-pan and warranted to retain its heat for five or six hours. A cold garret library was Goethe's favorite refuge from sick-headaches; and the Chevalier Edelkranz reminds his fur-loving countrymen that, when the difference of temperature between the external air and that within-doors is inconsiderable, it would be useful to "put on an extra coat on *returning* home, instead of doing it when going out, since the exercise in the open air produces the necessary degree of warmth, which, in the chamber, in a sedentary state, can only be supplied by additional clothing."

In our climate, however, there are days when a child of the Caucasian race has urgent need of all the overcoats his shoulders can support, and the natives of Northern Michigan have taught their Saxon neighbors some useful lessons in the art of surviving a Lake Superior snow-storm. Experience has made them eschew our common head-gear; they wear "Mackinaw hoods," a sort of monk's cowl, buttoned to the mantle-collar and covering every part of the face but the eyes and a small space between the mouth and the nostrils; double woolen mittens, reaching half-way up to the elbow; baggy trousers, fastened around the ankle, and shoes that admit three or four pairs of worsted stockings. Their particular care seems to be to protect the neck, hands, and feet; and it might, indeed, be accepted as a general rule that the parts of the body farthest from the heart are most liable to suffer from the effects of a low temperature. All extremities—toes, fingers, nose, and ears—are especially apt to get frost-bitten, but marching against a cold wind also produces a peculiarly uncomfortable sensation about the *neck*, and I can not help thinking that there is something wrong about our fashion of cropping our boys like criminals. A good head of hair may be something more than an ornamental appendage, and Nature seems to have taken especial care to protect the nape of the neck in a great number of different animals. It is certainly a suggestive circumstance that fomenting the space between the shoulders exerts an assuaging effect on various affections of the respiratory organs; and, if I had the care of a boy with an hereditary disposition to a pulmonary disease, I should feel strongly tempted to defy fashion, and let him wear his hair *à la Guido*—about a foot long.

The canal-laborers of Sault Ste. Marie wear double hoods, and on many days have to stuff them with wool to save their ears; but, in the more populous part of America, such days are a rare exception, and south of the lower lakes the average school-boy will prefer to rough it with a tippet shawl or a common cap with a pair of ear-flaps. In regard to the utility of woolen underclothes, opinions are much divided: Carl Bock recommends worsted jackets; Dr. Coale flannel undershirts and drawers, with extra breast-pads in cold weather; but the hardy Scandinavians, Russians, and French Canadians, as well as the great majority of our German population, still stick to coarse linen next the skin, and use woolen pectorals only as counter-irritants in rheumatic affections. Persons who can not bear woolen underclothes, I would advise to try the Normandy plan of ruffled linen, which might be applied even to hosiery and drawers. Chamois-leather, too, is as warm as wool and less irritating to the skin, and has the advantage of being more durable, and withal cleanlier, than the best flannel. On stormy days, especially during the piercing northwest storms of our prairie States, few children will object to a Scotch plaid, worn like a burnoose, over head and shoulders, or a handful of wool stuffed around the socks in a pair of wide brogans.

But at the beginning of the warm season all such things ought to be thrown aside. A loose shirt, linen jacket, and short linen trousers are the right summer dress for a healthy boy—a dalmatica and light straw hat for a healthy girl—in a country where the six warmest months approach the isotherms of Southern Spain. No wadded coats, no drawers, and, in the name of reason, no flannels, nor shoes and stockings, unless the mud is very deep, or the road to school recently macadamized. The long-lived races of Eastern Europe would laugh at the idea that the constitution of a normal human being could be endangered by an April shower, or that in the dog-days "health and decency" require a woolen cuticle from neck to foot. Have dogmas and hearsays entirely closed our senses to the language of instinct, to the meaning of the discomfort, the distracting uneasiness under the burden of a load of calorific covers and bandages, while every pore of our skin cries out for relief, for the cooling influence of the free open air? Keep your children under lock and key, lest the sun should spoil their complexion or their morals, let them pass their days in an under-ground dungeon like Kaspar Hauser, but do not load them with woolen trappings at a time when even a linen robe becomes a Nessus-shirt. There is a story of a glutton being cured by a friend who persuaded him to eat and drink nothing for twenty-four hours without putting an equivalent in quantity and quality into an earthern crock, and the next day made him inspect the *collectanea;* and on the same principle a person of common sense might perhaps be redeemed from the slavery of the dress-mania, by making him wrap up his complete suit of traps and weigh the bundle: he would find that the summer dress of a fashionable gentleman outweighs the winter coat of the most hirsute brute of the wilderness. A grizzly bear, shorn to the skin, would yield about ten pounds of hair and wool; but a dandy's accoutrements—flannel undershirt, drawers, shoes, stockings, starched overshirt, waistcoat, cravat, black dress-coat, and pantaloons—would weigh at least fourteen pounds. Habit mitigates the evil, though there are times when the martyrs of fashion suffer more in a single hour than a ragged Comanche in the coldest winter week; but, for boys and young girls, calorific food and woolen clothes certainly make the sunniest days the saddest in the year.

The vicissitudes of the weather? It is worth a journey to Trieste to see the youngsters of the suburbs enjoy their evenings on the Capo Liddo, the sandy headland between the Pola pike-road and the harbor fortifications; four or five hundred half-wild boys, splashing in the surf, throwing stones, wrestling, or chasing each other along the shore, all shouting and cheering, merry as carnivallers, though there is not a pair of shoes or a dozen hats in the crowd. Swift-footed, lithe, and indefatigable, they are the very picture of careless health; you can see them at play almost every evening, even in winter, when the *Tramontane* raises the snow-drifts of the Karst. They laugh at summer showers; their linen jackets will dry before they get home. Sunshine makes them a holiday; but let your well-dressed New York or Paris school-boy join in their sports, and examine his clothes after an hour or two, and see if perspiration has not made his undershirts as wet as any rain could make his jacket.

Decency? Are the gambols of a barefoot boy more unseemly than the contortions of a sun-struck alderman in his holiday dress? Can ethics or æsthetics be promoted by the imprecations of a sleepless victim of flannel night-shirts and closed bedroom windows? If daily misery can spoil the temper of a saint, the ladies of the American Dress-Reform are working in the interest of charity and good-humor by removing a chief incentive to the opposite sentiments, for the aggravations of Tantalus must have been trifling compared with those of an American school-girl *à la mode*, at the thought of a mountain meadow to run on with naked feet or a shady brook to pick pebbles from with bared arms. Pocahontas, indeed, had no need to envy the "fair maids in the land of her lover," if the fair ones had to wear the twenty-three distinct pieces of dry-goods which, according to a correspondent of Virchow's "Jahresberichte," constitute the summer dress of the average girl of the period. The blind submission to such demands of fashion can be explained only by a long subjection of human reason to authority, together with that ridiculous *dread of nudity* which forms a characteristic feature of all anti-natural religions. According to the ethics of the Hebrew-Buddhistic moralists, all *naturalia sunt turpia;* the body is the arch-enemy of the soul, and must be hidden, lest the children of the Church might be reminded of their relationship to the despised children of Nature. Boys and girls have no vote in such matters, or they would consent to turn night into day for the sake of getting a little exercise without the dire alternative of sweating to death or awakening the anathemas of Mrs. Grundy. The misery reaches its climax in June, when the warm weather begins before the vacations; and in midsummer a person with humane instincts would rather make a wide *détour* than pass a town school or a cotton-factory and witness the triumph of our pious civilization—the daily and intolerable torture of thousands of helpless children to please an Old Hypocrites' Christian Association of priests and prudes!

As houses have been called exterior garments, a heavy suit of clothes might be called a portable house —a protective barrier between the skin and the cold air; but in warm weather the most effectual device for diminishing the benefit of out-door exercise. Between May and October man has to wear clothes enough to keep the flies and gnats from troubling him: a pair of linen trousers, a shirt, and a light neckerchief—whatsoever is more than these is of evil. The best head-dress for summer is our natural hair; the next best a light straw hat, with a perforated crown. Hats and caps, as a protection from the vicissitudes of the atmosphere, are a comparatively recent invention. The Syrians, Greeks, Romans, Normans, and Visigoths wore helmets in war, but went uncovered in time of peace, in the coldest and most stormy seasons; the Gauls and Egyptians always went bareheaded, even into battle, and a hundred years after the conquest of Egypt by Cambyses (b. c. 525), the sands of Pelusium still covered the well-preserved skulls of the native warriors, while those of the turbaned Persians had crumbled to the jaw-bones. The Emperor Hadrian traveled bareheaded from the icy Alps to the borders of Mesopotamia; the founders of several monastic orders interdicted all coverings for the head; during the reign of Henry VIII, boys and young men generally went with the head bare, and to the preservation of this old Saxon custom Sir John Sinclair[*]

[*] "Code of Health and Longevity," p. 298.

ascribes the remarkable health of the orphans of Queen's Hospital. The human skull is naturally better protected than that of any other warm-blooded animal, so that there seems little need of adding an artificial covering; and, as Dr. Adair observes, the most neglected children, street Arabs and young gypsies, are least liable to diseases, chiefly because they are not guarded from the access of fresh air by too many garments (Adair's "Medical Cautions," p. 389). It is also well known that baldness is the effect of effeminate habits as often as of dissipation; and yet there are parents who think it highly dangerous to let a boy go out bareheaded even in May or September. The trouble is, that so many of our latter-day health codes are framed by men who mistake the exigencies of their own decrepitude for the normal condition of mankind. Thousands of North American mothers get their hygienic oracles from the household notes of some orthodox weekly, where the Rev. Falstaff Tartuffe assures them —from personal experience—that raw apples are indigestible, and that rheumatism can be prevented only by night-caps and woolen undershirts.

Girls, it seems, have to pass through a millinery climacteric, as their brothers through a wild-oats period; but even during that interregnum of reason the instinct of self-preservation would assert its supremacy if the health laws of physiology and their antagonism to certain fashions were more generally understood. Claude Bernard speaks of a French philanthropist who proposed to offer a prize for the most tasteful female dress, manufactured from the cheapest materials; and, if the votaries of the Graces would consent to a reform in the shape and stuff of their garments, we could well afford to indulge them in chromatics and a flounce or two, for there is no reason to afflict them with Quaker-drab, if more cheerful colors are as cheap. As long as they avoid excesses in the quantity and form of their dress, and restrict themselves to four dimes' worth of vanities per month, we need not grudge them a display of their taste in the selection of pretty patterns; let them **radiate** in all **the colors of the rainbow** and all the gems of the "Chicago Prize-Package Company." *Veniunt a veste sagittæ*—the dress problem has always employed the leisure of gossips and Doctors' Commons, especially in cities, and more especially in the wealthy and indolent cities of the Old World. There is a legend of a New England virgin fainting at the mention of "undressed lumber," but that tradition must be of Eastern origin. The dry-goods worship is carried nowhere further than where children are treated like dolls and women like children, unfit to be intrusted with any more important business. The "organ of ornamentativeness," or fashion-**mania, may, after** all, not be an in**nate** instinct **of the female mind.** Madame de Staël and Mrs. Lewes at least deny it, and, if they are right, an enlarged sphere of activity will by-and-by help their sisters to outgrow that bias. In the mean while, the best palliative is a liberal education, besides a zealous propaganda of the two chief theses of the *dress reform:* wider **jackets** and shorter under-garments; no trailing dresses, **keeping the** feet wet and impeding locomotion; no stays, **corsets, and** strait-jacket bodices.

Next to **the regulation dress** of the Turner hall, the present style of the United States infantry uniform is about the most sensible that could be devised with regard to sanitary advantages, and nearly so in respect to good taste, if Thorwaldsen's dictum holds good, that the **most becoming** garments are those which adapt themselves to the natural outlines of the human form. A jacket should be loose, with wide but rather short sleeves, loose trousers, no waistcoat or drawers in the summer season; for small boys, short trousers without pockets, but with broad leather braids along the seams. The comparative advantages of waistbands or braces have been frequently controverted; at best it is only a question of choosing the lesser evil. A tight belt is almost as injurious as a corset, while non-elastic suspenders may interfere with the functions of the respiratory organs, and even occasion stooping. For boys and slender-built men, with well-developed hips, an elastic waistband is, on the whole, preferable; corpulent persons can not dispense with braces, for the plan of buttoning the breeches to the jacket or waistband would amount to the same, by making the shoulders support the weight of the lower garments. Tight breeches have, fortunately, gone out of fashion; likewise tight kid-gloves, which were once *de rigueur* on every public promenade.

But we all sin against our feet; not one white man in ten thousand wears shoes that are not more or less of a hindrance in walking, and often a source of wretched discomfort. In the United States, England, and Central Europe, it is wholly impossible to find a ready-made pair of shoes to fit a normal human foot; they are all too tight in proportion to their length, every pair of them, even the United States army shoes and the English "fast-walking brogans." Heels are nonsense; there is no excrescence on the sole of a well-formed human being. A man can walk faster, more easily, and more gracefully, with level shoes, with soles shaped like those of a slipper or an Indian moccasin. An easy shoe should be heelless; the upper leather soft and pliable; the sole of a No. 9 shoe at least four inches wide. But you can not persuade a shoemaker to commit such heresies against the tenets of his craft. Dio Lewis recommends paper patterns, corresponding to the exact shape of the natural sole, but it is all in vain; a compromise between reason and dogma is the best you can attain by such means. The only practicable plan is to get one pair of shoes made under your personal supervision, and then stipulate for the necessary number of precise *fac-similes*. The disciple of St. Crispin shrinks from the guilt of the original sin, but connives at a copy; a precedent will reconcile his conscience.

For children there is a shorter expedient: let them go barefoot, at least in-doors and all summer; it will make them hardier and healthier. Abernethy, Schrodt, Dr. Adair, Jean Jacques Rousseau, and Claude Bernard, agree on this point; Dr. Cadogan thinks shoes and stockings wholly useless, and John G. Whittier seems to share his opinion that a barefoot boy is the happiest representative of the human species. "I can see no reason why my pupil should always have a piece of ox-hide under his foot," says the author of "Émile." . . . "Let him run barefoot wherever he pleases. . . . Far from growling about it, I shall imitate his example." *

Refusing to buy tight shoes might bring easy ones into fashion; but boys are better off without them, especially in the years of rapid growth, when their

---

* "Pourquoi faut-il que mon élève soit forcé d'avoir toujours sous les pieds une peau de bœuf? Quel mal y aurait-il que la sienne propre pût au besoin lui servir de semelle? Il est clair qu'en cette partie la délicatesse de la peau ne peut jamais être utile à rien et peut souvent beaucoup nuire. Que Émile coure les matins à pieds nus, en toute saison, par le chambre, par l'escalier, par le jardin; loin de l'en grondir je l'imiterai."—(Rousseau: "Émile, ou de l'Éducation," p. 143.)

measure changes from month to month, for too wide shoes are as uncomfortable as tight ones. Out-doors, children's stockings are almost sure to get wet, and keep the feet clammy and cold; while a young gypsy or a Scotchman, inured to wind and weather, treads with his bare feet the swampiest valleys and the roughest hill-roads without the least discomfort. Nature produces a better sole-leather than any shoemaker; the tegument of a raccoon's foot or a monkey's hind-hand can give us an idea of the marvels of her workmanship. The sole of a plantigrade animal is not hard; on the contrary, quite pliable and soft to the touch, but withal *tougher* than any caoutchouc, impervious alike to water, sand, and thorns. A camel, too, has a foot of that sort —pads that resist the burning gravel of the desert for years, where a horse's hoof would wear out in a few weeks; for the same reason that a "sand-blast" destroys tanned sole-leather and horn, but hardly affects the elastic skin of the human hand. Millions of unshod Hindoos, negroes, and South American savages brave the jungles of the tropical virgin woods; and in Nicaragua I saw two Indian mail-carriers *trot* barefoot over the lava-beds of Amilpas, over fields of obsidian and scoria, where a dandy in patent-leather gaiters would have feared to tread. Three or four seasons of barefoot rambles over the fields and hills will develop such soles—natural shoe-leather that improves from year to year, till it can be warranted to protect the wearer against the roughest roads, and, as the experience of our half-wild frontiersmen attests, also against colds and rheumatism. A mere moccasin secures such hardy feet against frost-bites; for here, too, the rule holds good that those who keep themselves too warm in the summer season deprive themselves of the advantage to be derived from additional clothing in cold weather and in old age.

Herr Teufelsdröckh devoted a voluminous work to the "Philosophy of Clothing," but the practical part of the science may be summed up in a few words. Our dress ought to be adapted to the changes of the seasons, and should be in quality durable, cleanly, and, above all, easy; in quantity, the least amount compatible with decency and comfort.

CHAPTER VI.

SLEEP.

"Children, stinted in their sleep, are never wide-awake."—PESTALOZZI.

THE vital processes of man, like those of all his fellow-creatures, are partly controlled by automatic tendencies. Some functions of our internal economy are too important to be trusted to the caprices of human volition; breathing, eating, drinking, and even love, are only semi-voluntary actions; and during a period varying from one fourth to two fifths of each solar day the conscious activity of the senses undergoes a complete suspense: the cerebral workshop is closed for repairs, and the abused or exhausted body commits its organism into the healing hands of Nature. Under favorable conditions eight hours of undisturbed sleep would almost suffice to counteract the physiological mischief of the sixteen waking hours. During sleep the organ of consciousness is at rest, and the energies of the system seem to be concentrated on the function of nutrition and the renewal of the vital energy in general; sleep promotes digestion, repairs the waste of the muscular tissue, favors the process of cutaneous excretion, and renews the vigor of the mental faculties.

The amount of sleep required by man is generally proportionate to the waste of vital strength, whether by muscular exertion, mental activity (or emotion), or by the process of rapid assimilation, as during the first years of growth and during the recovery from an exhausting disease. The weight of a new-born child increases more rapidly than that of a eupeptic adult, enjoying a liberal diet after a period of starvation, and, though an infant is incapable of forming abstract ideas, we need not doubt that the variety of new and bewildering impressions must overtask its little sensorium in a few hours. Nurslings should therefore be permitted to sleep to their full satisfaction; weakly babies, especially, need sleep more than food, and it is the safest plan never to disturb a child's slumber while the regularity of his breathing indicates the healthfulness of his repose; there is little danger of his "oversleeping" himself in a moderately warmed, well-ventilated room. Never mind about meal times: hunger will awaken him at the right moment, or teach him to make up for lost time. Three or four nursings in the twenty-four hours are enough; Dr. C. E. Page, who has made the problem of infant diet his special study, believes that fifty per cent of the enormous number of children dying under two years of age are killed by being coaxed to guzzle till they are hopelessly diseased with fatty degeneration.*

The healthfulness of village-children is partly due to the tranquillity of their slumber in the comfortable nooks of a quiet homestead, or in the shade of a leafy tree, while their parents are at work in a way rather

---

* "The only wonder is that any infant lives sixty days from birth. Fed before birth but three times a day, he is after birth subjected to ten or twenty meals in the twenty-four hours, until chronic dyspepsia or some acute disease interferes. . . . So far from admitting a possible error in advising three meals only, I am convinced that, for a hand-fed baby especially, two would often be better than three."—("How to feed a Baby to make it healthy and happy," p. 55.)

incompatible with the habit of fondling the baby all night. In houses where there is plenty of room, the nursery and the infant's dormitory ought to be two separate apartments: the play-room can not be too sunny; for the bedroom a shady and sequestered location is, on the whole, preferable. Next to out-door exercise, silence and a subdued light are the best hypnotics. But under no circumstances should insomnia be overcome by cradling or narcotics. Stupefaction is not slumber. The lethargy induced by rocking and cradling is akin to the drowsy torpor of a seasick passenger, and the opium-doctor might as well benumb his patient by a whack on the head. The morbid sleeplessness of children may be owing to several causes which can be generally recognized by the symptoms of their *modus operandi;* impatient turning from side to side, as if in a vain attempt to obtain a much-needed repose, means that the room is too stuffy or too warm; long wakefulness, combined with squalling-fits and petulant movements, indicates acidity in the stomach (overfeeding, or too much "soothing-sirup")—let the little kicker exercise his muscle on the floor; in malignant cases, skip a meal or two, or give water instead of milk. After weathering an attack of croup, children often lie motionless on their backs with a peculiar glassy stare of their wide-open eyes. Leave them alone; instinct teaches them to assuage the distress of their lungs by slow and deep respirations; rest and a half-open window will do them more good than medicine.

Healthful infants—i. e., under rational management the great plurality—can soon be taught to transact their public business at seasonable hours, or at least to abstain from midnight serenades. If mothers would make it a rule to do all their nursing and fondling in the day-time, their little revivalists would soon learn to associate darkness with the idea of silence and slumber. Habit will do wonders in such things. Captain Barclay and several American pedestrians learned to take their half-hour naps as a traveler snatches a hasty lunch, and many old soldiers develop a faculty of going off to sleep, as it were, at the word of command, the moment their shoulders touch the guard-house bunk. The two drowsiest years of my life I passed at an old-style boarding-school, where teachers and pupils were limited to seven hours of sleep, after nine hours of study, besides written exercises and special recitations, and where sixty or seventy of us had to sleep in a large hall; and I do not believe that the last flickering of our five-minutes candle was ever witnessed by a pair of more than half-open eyes.

But that same faculty of sleeping and waking at short notice may be utilized for the purpose of taking little naps whenever opportunity offers—in the last half-hour of the noontide recess, or during the Buncombe interacts of a protracted session. The inhabitants of all intertropical countries make the time of repose a movable festival, and during the dog-days of our torrid summers it would clearly be the best plan to imitate their example. "Children must not sleep in the day-time," says a by-law of our time-dishonored Koran of domestic superstitions; and, not satisfied with keeping our little ones at school during the drowsy afternoons of the summer solstice, we increase their misery by stuffing them at the very noon of the hottest hours with a mass of greasy (i. e., heat-producing and soporific) food. An hour after the end of a long, sultry day comes the cool night-wind, heaven's own blessing for all who hunger and thirst after fresh air; but no, "Night-air is injurious"; besides, Mrs. Grundy objects to promenades after dark, so the children are driven to their suffocating, unventilated bedrooms, not to sleep but to swelter, till toward midnight, when drowsiness subsides into a sort of lethargy which yields only to broad daylight, three or four hours after sunrise. "So much the better," says the fashionable mother, who has passed the night at an ice-cream *ridotto*, "and morning air isn't healthy, either; most dangerous to leave the house before the dew is off the grass."

Only the curse of pessimism, our woful distrust of our natural instincts, can explain such absurdities. The parched palate's petition for a cooling liquid is not plainer than the brain's craving for rest and slumber when a high temperature adds its somniferous tendency to the drowsy influence of a full meal. On warm summer days all Nature indulges in a noontide nap; I have walked through tropical forests that were as silent under the rays of a vertical sun as a Norwegian pine-grove in the dead of a polar night; nor would it be easy to name a single animal that does not appear sleepy after meals. At noon leaf-trees throw their densest shade; even butterflies seek the penetralia of the foliage, and lizards cling lazily to the dark side of the lower branches; every school-teacher knows that children feel the drowsy spell of the afternoon sun; why should they alone be hurt by yielding to its promptings? Either postpone the principal meal to the end of the day, or increase the noontide recess to at least three hours, so as to leave time for a digestive *siesta*.

In midsummer all mammals (squirrels, perhaps, excepted) become semi-nocturnal: deer and llamas pasture the moonlit mountain-meadows; bears, badgers, and the larger species of monkeys are wide-awake; buffaloes wander *en masse* to the next drinking-place; and the step-children of Nature, the starved lazzaroni of Southern Europe, forget their misery if they can procure a fiddle or a guitar. The moonlit streets of the Mexican cities swarm with merry children, but north of the Rio Grande not a decent lad is seen out-doors after sundown; Luna has to seek her Endymions in the tropics, though our summer nights are often as glorious as the *noches serenas* of Southern Andalusia. And what would our hardy forefathers have said about our dread of the morning dew? How many thousands of hunters and soldiers have slept in the open fields, and how many times did we *wade* through the dew-drenched brambles of the Ardennes, my little brother and I, to see the sun rise, and breathe the mountain wind, at the only hour when the air is both fragrant and cool, inspiring thoughts which music can only awaken for a fleeting moment!— if such hours are mortiferous, there can not be a more agreeable way of ending what our latter-day epicures are pleased to call life.

What harm can there be in dividing our daily portion of sleep? Birds and beasts do it, the founders of the most ascetic orders of Spanish monks allowed it, and our summer months are certainly as warm as those of Southern Europe. People who are so anxious to improve the shining hours for business purposes had much better curtail the number of their meals; take a vote among the juvenile operatives of a cotton-factory, and ten to one that a large majority would gladly postpone, or even renounce, their dinner for the privilege of sleeping an hour or two between 1 and 3 P. M. A Belgian silk-manufacturer, who had spent his own boyhood at the loom, told me that he could never find it in his heart to discharge a factory-child for dozing over its work.

Necessity may compel individuals to compromise such matters. If I had to work or teach all day, I would not eat a crumb between breakfast and supper, and pass the dinner-hour under a shade-tree; but parents who can afford to educate their children at home should give them either an all-summer vacation or a half-afternoon recess—let them rest from twelve till three, or sleep if they prefer; in the evening, do not send them to bed till they are really tired, and till the night-wind has revitalized the air of their bedrooms; but make them rise with the sun—if they are drowsy they will go to bed earlier the next evening. There is no danger of a child's—especially a boy's—oversleeping himself, unless the hardships of his waking hours are so intolerable that oblivion becomes a blessing; but it can do no harm to make the health-giving morning hour as attractive as possible: provide some out-door amusement, a prize foot-race, a butterfly-hunt, or gathering windfalls in the apple-orchard; if the desire for longer sleep can outweigh such inducements, there must be something wrong—plethorific diet, probably, or over-study. The requisite amount of sleep depends on temperament and occupation as well as on age; with children under ten, however, too much indulgence would be an error on the safer side: let them choose their allowance between eight and ten hours; in after-years, seven hours should be the minimum, nine the maximum for *healthy* children; sickly ones ought to have *carte blanche*, both as to quantum and time of repose; consumptives, especially, need all the rest they can get. Profound sleep in a cool, quiet retreat is Nature's own specific for all wasting diseases, a panacea without price and money.

Nothing can be more injudicious than to stint children in their sleep with a view of gaining a few hours for study. "That plan," says Pestalozzi, "defeats its own purpose, for such children are never wide-awake; you can keep them out of bed, but you can not prevent them from dozing with their eyes open. A wide-awake boy will learn more in one hour than a day-dreamer in ten."

Habitual deficiency of sleep will undermine the strongest constitution; headache, throbbing, and feverish heat are the precursors of graver evils, unless a temporary loss of mental power compels an armistice with outraged Nature. King Alfred, Spinoza, Kepler, Victor Alfieri, Madame de Staël, and Frederick Schiller killed themselves with restless study; Beethoven and Charles Dickens, too, probably prepaid the debt of Nature by their habit of fighting fatigue with strong coffee. Sleeplessness may lead to chronic hypochondria, and even to idiocy; without their long vigils, the monks of the Thebais and the fathers of the Alexandrian Church could hardly have written such stupendous nonsense. It is a curious fact that compulsory wakefulness combined with mental activity often induces a state of morbid insomnia, an absolute inability to obtain the sleep which it was at first so difficult to resist. In such cases, the only remedy is fresh air and a complete change of occupation. During sleep the brain is in a comparatively bloodless condition;* a hot head and

---

* Dr. Caldwell records a case of a woman at Montpellier, who "had lost part of her skull (from disease), the brain and part of its membranes lying bare. When she was in a deep and sound sleep, the brain lay in the skull almost motionless; when she was dreaming, it became elevated;

throbbing temples are unfavorable to repose, and it has been suggested that insomnia might be counteracted by a hot foot-bath, chafing the arms and legs, or any similar operation that would divert the blood from the head toward the extremities, and thus tend to diminish the activity of the cerebral circulation. Listening to distant music or the ripple of a river-current has also a wonderful hypnotic effect, the repetition of monotonous sounds, or, indeed, of any sensorial impression, seems more favorable to repose than their entire absence. The philosopher Kant assures us that he could obtain sleep in a paroxysm of gout by resolutely fixing his attention on some abstruse ethical or mathematical problem, but remarks that the success of that method depends upon the laboriousness of the mental process; the mind, as it were, takes refuge in sleep as the alternative of drudging at a wearisome task. Robert Burton, too, gives a number of similar recipes, besides a list of wondrous medicinal compounds to be swallowed or inhaled *ad horam somni*, but in ordinary cases it is better to try the effects of out-door exercise, before resorting to dormouse-fat,* theological text-books, or other desperate remedies.

Being naturally a sound and long sleeper has been ranked among the surest prognostics of a long life, and sleep after a wasting disease as the most certain symptom of recovery. Most brain-workers are subject to occasional fits of insomnia, but the faculty of sustaining

and, when she awoke, it became suffused with blood and seemed inclined to rise through the cranial aperture."—("Psychological Journal," vol. v, p. 74.)

* "Anoint the soles of the feet with the fat of a dormouse, the teeth with ear-wax of a dog, swine's gall, oil of nunaphar, henbane," etc.— ("Correctors of Accidents to procure Sleep," "Anatomy of Melancholy," p. 414.)

health and vigor upon a very small allowance of sleep is generally a concomitant of mental inferiority, or at least inactivity. The most intelligent animals, dogs and monkeys, sleep the longest; stupid brutes merely stretch their legs, their inert brain requires no rest; a cow never sleeps, in the proper sense of the word. Mirabeau, Goethe, and James Quin often slumbered for twelve or fourteen hours successively, while Leopold I, of Austria, and Charles IV, of Spain, the heartless and brainless bigots, could content themselves with five hours of sleep out of the twenty-four, and their prototype, the Emperor Justinian, often even with *one*.—(Gibbon's "Rome," vol. vii, p. 406.)

Heinrich Heine wonders why Jehovah did not square his account with our wicked forefathers by punishing them in their sleep, instead of compromising their innocent progeny. Dietetic sins often avenge themselves in that way; blutwurst, sauerkraut, or short-cakes with pork-fritters, generally result in apocalyptic visions, and an eel-pie for supper is a reliable receipt for a first-class nightmare. Vivid dreams, *per se*, however, are by no means a morbid symptom; on the contrary, the scenes of the slumber-drama are most lively and life-like in the happiest years of childhood; and I remember a time when I longed for the bed-hour, in anticipation of a pleasant dream-land excursion. Children are apt to relate their trance adventures, and I would encourage the habit; dreams, as the elder Pliny already observes, may often afford a suggestive insight into the ethical condition of the mind; also into the condition of the stomach. Melodramatic incidents indicate the presence of irritating ingesta, and the least attempt at clairvoyance calls for out-door exercise and an aperient diet. A South-German feather-bed is a Trophonian cave; the difficulty of turning from side to side crowds the brain with alarming phantasms, and the excessive warmth of the thing itself is apt to affect the imagination. The best bed is, indeed, a hard, broad mattress, or a well-stuffed straw tick, and, for reasons I have stated in the chapter on "In-door Life," a woolen blanket over a linen bed-sheet is preferable to a quilt. Those who find it uncomfortable to sleep in an absolutely horizontal position should slightly raise the head-end of the bedstead rather than use a thick bolster. A thick pillow bends the head upon the breast, or keeps the neck in a position that aggravates the distress of respiratory difficulties. Woven-wire mattresses recommend themselves by their cleanliness and durability; their elastic qualities alone would hardly justify a great expense, though luxury has even devised an "hydrostatic bed," a trough of water with a tegument of caoutchouc. History records the name of the Sybarite who "cried aloud because a leaflet of his flower-mattress got crumpled"; and Chevalier Luckner, the Russian Lucullus, built himself an air-pillow bed on noiseless wheels, that could be turned by a hand-lever, in order to move the sleeping-car from or toward the stove, the aphelion and perihelion being determined by the state of the out-door atmosphere. Such chevaliers deserve the penance of Ezekiel (iv, 3–6), who had to lie three hundred and ninety days on his left side for the iniquity of the house of Israel, and forty days extra for the iniquity of the house of Judah. A weary head needs no air-cushions, with noiseless wheel-attachments; brakesmen take their intermittent naps on the hard caboose-bunk of a rumbling freight-train; and the log of the Royal Sovereign records that, during the heat of the battle of the Nile, some of the over-fatigued boys fell asleep upon the deck.

The habit of going to sleep at fixed hours can overcome the tortures of neuralgia, and some practical stoics have manifested a not less astonishing command over their mental emotions; **Napoleon I** slept soundly on the eve of the **battle he knew to be his** last chance, like Mohammed II **before his last** neck-or-nothing assault upon the ramparts of Constantinople. Army-life can acquaint a man with strange beds, as well as bedfellows. Skobeleff's troopers had to sleep in dug-outs on the woodless ridges of the Balkan; and, during Ney's retreat from Moscow, the commander himself had once to pass a night in a root-house, where a few rotten boards and a bundle of straw formed his only protection against a raging snow-storm.

But "roughing it" teaches some useful lessons, and soldiers and hunters often learn by experience that sleep under such circumstances depends upon the possibility of *getting the feet warm;* rain in the face, or even a wet overcoat, is less anti-hypnotic than chilled toes. In a trapper's bivouac the sleepers generally lie in a circle around the camp-fire, with their feet toward the glowing embers, and the Swiss mountaineers use foot-sacks —long socks of a felt-like stuff, and wide enough to leave room for a lot of dry leaves, besides two or three pairs of stockings. Both methods are practical applications of Dr. Caldwell's theory that a decrease of the cerebral blood-circulation has a somniferous influence; in other words, that sleep can be promoted by warming the extremities of the body, and thus diverting the blood from the head.

In-doors, summer often reverses the problem; in the dog-days, when the amount of bedclothing has to be reduced to a minimum, the main point is to cool the head by lowering the temperature of the bedroom.

Open windows, a hard, smooth mattress, linen bed-sheets, and a light supper will generally answer the purpose; in the lower latitudes, George Combe recommends glazed brick floors, frequent sprinklings, and in very hot nights a tub with ice. And why not? The Turkish residents of Damascus pass the summer nights in the *yeyirman* or fountain-hall of their cool houses, and the garrison soldiers of San Juan d'Ulloa deem it a special privilege to sleep on the floating wharf, exposed to the spray and the fitful swell of the Gulf-tide.

In the West Indies and the Mississippi Valley, mosquito-bars are a sad necessity, but all sensible people should be glad that the French canopy-beds are going out of fashion. The French are right, though, in making children over ten years sleep alone; it is one of the rare instances of an etiquette law being supported by a valid reason. To those who can afford it, Dr. Franklin recommends even two beds per individual, and in sweltering summer nights it is certainly a blessing to be able to leave a hot bed for a cool one; in the large family guest-chambers of a German hotel, sleepless travelers can thus change the beds like relay-horses. The builders of the old English country-seats seem to have made it a rule to have the houses face due south, with few or no windows on the north side, and in such buildings the east windows would make the best bedroom fronts, both on account of the evening shade and the monitory morning sun. In our Northwestern Territories, where the thermometer ranges from 90° above zero to 45° below, it would be no bad plan to vary the location of the bed-chamber with the change of the season, but, as a general rule, the dormitory should be the coolest room in the house—i. e., the nearest to the north side, and the farthest from the kitchen.

CHAPTER VII.
RECREATION.

"Mirth is a remedy."—Thomas Hobbes.

Happiness is the normal condition of every living creature, for in a state of nature every normal function is connected with a pleasurable sensation. "To enjoy is to obey"; if human life were what it could be and what its Author intended it to be, the path of duty would be a flowery path, the reward of virtue would not be a crown of thorns; man, like all his fellow-creatures, would attain to his highest well-being by simply following the promptings of his instincts. Wild animals have not lost their earthly paradise; he who has observed them in the freedom of their forest homes can not doubt that to them existence is a blessing, and death merely the later or earlier evening of a happy day. Nor would our missionaries find it easy to persuade an able-bodied savage that earth is a vale of tears, till fire-water and fire-arms demonstrate the superiority of revelation over the light of nature. The children of the wilderness need no holidays; to them life itself is a festival and earth a play-ground for manifold games, not the less entertaining for being sometimes spiced with danger or prompted by hunger and thirst.

But in process of time the daily life of a combatant in the harder and harder struggle for existence became so joyless and wearisome that the clamors of an unsatisfied instinct suggested the institution of periodical festivals: pleasure-days intended to offset the tedium of monotonous toil, as gymnastic exercises tend to counteract the influence of sedentary occupations. The Assyrians and Greeks had tri-monthly holidays, besides annual revels, and great national festivals at longer intervals. In ancient Etruria every new month was ushered in by a day of merry-making in honor of a tutelary deity; the patricians and plebeians of republican Rome had their field-days; the festivals of the seasons united the pleasure-seekers of all classes, and even the slaves had their Saturnalia weeks when some of their privileges were only limited by their capacity of enjoyment. In the first centuries of the Roman Empire, when the growth of the cities and the scarcity of game began to circumscribe the private pastimes of the poorer classes, the rulers themselves provided the means of public amusements; at the death of Septimus Severus (A. D. 211), the capital alone had six free amphitheatres and twelve or fourteen large public baths, where the poorest were admitted gratis, and none but the poorest could complain about the half-cent entrance-fee to the luxurious *thermæ*. The *circenses*, or public games, were by no means confined to the gladiatorial combats that have exercised the eloquence of our Christian moralists; dramatic entertainments, trials of strength, and the exhibition of outlandish curiosities, seem to have been as popular as the grandest prize-fights, unless the combatants were international champions. And it would be a great mistake to suppose that only the wealthy capital could afford to amuse its citizens at the public expense; from Gaul to Syria every town had a circus or two, every larger village an arena, a free bath, and a public gymnasium. The Colosseum of Vespasian seated eighty thousand spectators, but was rivaled by the amphitheatres of Narbonne, Syracuse, Antioch, Berytus, and Thessalonica.* Children, married women, old men, and many trades-unions had their yearly carnivals, and, during the celebration of the Olympian and Capitoline games and various local festivals, even strangers enjoyed the freedom of the larger towns.

And now?—Professor Wirgmann, in his "Annalen des Russischen Reiches," estimates that since the accession of Nicholas I the modern Cæsars have expended an average annual sum of seventeen million dollars for the torture of their subjects; how many cents have they ever spent for national pastimes? How many spectators (since the abolition of the "Tyburn-days") have ever been entertained at the expense of the wealthy British Empire? What has our Great Republic done in the matter of *circenses*, except to pass an occasional sabbath law for the suppression of public amusements on the only day in which a large plurality of our working-men find their only leisure for recreation? The spoils of a Roman consul would dwindle before the rents of our American, German, and French financiers: what have our commercial triumphators ever achieved for the entertainment of their poor fellow-citizens? Cooper Institute lectures, street revivals, and prize distributions at the examination of a sabbath-school for adults? "At the proposition of such-like pastime," says Ludwig Boerne, "a resurrected citizen of ancient Rome would feel like a filibuster at an invitation to dive for copper coins in a duck-pond, after having chased King Philip's silver fleet on the Spanish Main."

Not poverty makes our daily ways so trite and joyless, for the best recreations are still as free as the air

* Tacitus, "Annalen," xli-xlv.

and the sea; nor want of leisure, for we manage to find plenty of time for humdrum ceremonies. The old Egyptians turned their funerals into holidays—we celebrate our holidays like funerals; all the employments of our weekly day of rest are sicklied over with a cast of superstitious fear; and, indeed, no other anachronism of our strangely complex civilization proclaims more loudly the necessity of its divorce from the influence of an anti-natural religion. When that religion reigned supreme, its exponents openly and violently waged war upon all earthly joys; sublunary life, according to their doctrine, was a state of probation for testing a man's power of self-denial; earth was the devil's own, and delight in its pleasures an insult to the jealous ruler of a higher sphere. They believed that God delights in the self-abasement and mortification of his creatures, and hoped to gain his favor by afflicting themselves in every possible way—by voluntary seclusion, fasts, vigils, the wearing of dingy garments, and abstinence from every physical pleasure. Failing to enamor mankind with their doleful heaven, they revenged themselves by depriving them of their earthly joys. In hopes of making the hereafter more attractive, they made life as repulsive as possible; kill-joys and persecutors were the active heroes of those times; ascetics and self-tormentors their passive exemplars. Virtue and joylessness became synonyms; men aspiring to superior merit exchanged the glories of the sunny earth for the misery of a gloomy convent; a "Man of Sorrows" became a type of moral perfection, an instrument of torture, the trade-mark of the new religion. *Kosmos*—i. e., beauty and harmony—was the oldest Grecian term for God's wonderful world; a "vale of tears" the favorite Christian epithet. A symposium of festive heroes was exchanged for a conventicle of whining penitents, Olympus for a charnel-house, the festival of the seasons for the ecclesiastic sabbath: there, a merry multitude, joining in dances and heroic games, inspired by the rapture of emulation, the joy of exuberant health and the beauty of earth till their happiness overflowed in anthems of praise to the bounteous gods; here, a cowed and wretched assemblage, listening with groans to the denunciations of a Nature-hating fanatic. And that hideous superstition founds its claim to our gratitude on its merit of having suppressed a few profligate pastimes—in aiming its death-blows at all earthly joys whatever; as if the crushing of a few poison-plants could atone for the attempt to turn a fertile continent into a sand-waste! The attempt, I say, for I do not believe that either the axe or the cross will for ever mar the beauty of our Mother Earth; the devastated woodlands of the East will ultimately be reclaimed, and here and there the moral desert of asceticism has already begun to bloom with flowers from the revived seeds of Grecian civilization.

Monachism, at least, is fast disappearing; in this age of railroads and steam-engines we have no time for positive self-torture *à la Simon Stylites*. But our commercial Pecksniffs have found it a time- and money-saving plan to stick to the negative part of the anti-pleasure dogma, and hope to atone for the dreary materialism of our daily factory-life by the still drearier asceticism of a Puritan sabbath: six days of misery in the name of Mammon, balanced by one day of sixfold misery in the name of Christ. "Worldly pleasures" are still under the ban of our spiritual purists; daily drudgery and daily self-denial are still considered the proper sphere of a law-abiding citizen, and special afflictions a special sign of divine favor. Life has become a socage-duty; we do not think it necessary to alleviate the distress of our poor till it reaches a degree that threatens to end it. We have countless benevolent institutions for the prevention of outright death, not one benevolent enough to make life worth living. Infanticide is now far more rigorously punished than in old times; we enforce every child's right to live and become a humble, tithe-paying Christian, but as for its claim to live happy we refer it to the sweet by-and-by. We shudder at the barbarity of the Cæsars, who permitted the combat of men with wild beasts, to cater to the amusement of the Roman populace; but we contemplate with great equanimity the misery of millions of our fellow-citizens, wearing away their lives in workshops and factories; millions of children of our own nation and country, who have no recreation but sleep, no hope but oblivion, to whom the morning sun brings the summons of a task-master and the summer season nothing but lengthened hours of weary toil; nay, we make it the boast of our pious civilization to deprive them of their sole day of leisure, to interdict their harmless sports, lest the noise, or even the rumor of their merriment, might disturb the solemnity of an assemblage of whining bigots. Hence the recklessness, the Nihilism, and the weary pessimism of our times, the melancholy that everywhere underlies the glittering varnish of our social life. Hence also that vague yearning after a happy hereafter, which the murderers of the happy past have made the principal source of their revenues.

With few exceptions the children of Christendom are stricken with a disease which mirth alone can cure. In North America and North Britain, especially, it is pitiful to witness the slow withering of so many light-loving creatures in the hopeless night of poverty and sabbatarianism; more pitiful to see the reviving of their spirits at every deceptive sign of dawn, the expedients of poor, compromising Nature, her make-shifts with half-recreations and half-sufficient rest, in the lingering hope of a better future—to come only with the repose from which no factory-bell can awaken a sleeper, when after long years of waning life, waning at last to a state of callous vegetation, Nature is reduced to the alternative of ending an evil for which she has no remedy.

But, while the ebb of life alternates with a tide, the struggle against a natural instinct is the struggle of Prometheus against the vulture of Jove; in the intervals of torment the martyr may forget his misery, but the torturer returns, and the poisoned arrows of the interventor can bring only a temporary relief. Man cannot conquer a God-sent instinct, though he may for a time defy it—with poison; the most incurable victims of intemperance are those who resort to stimulants less for the sake of intoxication than for the benumbing after-effect which helps them to stifle the voice of outraged Nature. It is a significant circumstance that the consumption of intoxicating poisons increases in times of famine and general distress; the Christian dogma of the reformatory value of misery has, indeed, been refuted by the most dreadful arguments of the world's history; the unhappiest nations are not only the most immoral but the most selfish and the meanest in every ugly sense of the word: virtues do not flourish on a trampled soil. The same with individuals; injustice, disappointment, and bodily pain, can turn the noblest man into a querulous tyrant, a harmless kitten into a spiteful cat. Happiness, on the other hand, is the sunshine that decks the moral world with flowers; making earth a heaven would be the surest way of turning men into angels; the hardest heart will melt under the persistent rays of kindness and happiness. Happy children have no time to be wicked; it is not worth their while to waste the merry hours on vices. Genius, too, is a child of light; the Grecian worship of joy favored the development of every human science, while the monastic worship of sorrow produced nothing but monsters and chimeras; for to modern science Christianity bears about the same relation as the plague to the quarantine.

But, aside from all this, mirth has an hygienic value that can hardly be overrated while our social life remains what the slavery of vices and dogmas has made it. Joy has been called the sunshine of the heart, yet the same sun that calls forth the flowers of a plant is also needed to expand its leaves and ripen its fruits; and without the stimulus of exhilarating pastimes perfect bodily health is as impossible as moral and mental vigor. And, as sure as a succession of uniform crops will exhaust the best soil, the daily repetition of a monotonous occupation will wear out the best man. Body and mind require an occasional change of employment, or else a liberal supply of fertilizing recreations, and this requirement is a factor whose omission often foils the arithmetic of our political economists.

To the creatures of the wilderness affliction comes generally in the form of impending danger—famine or persistent persecution; and under such circumstances the modifications of the vital process seem to operate against its long continuance; well-wishing Nature sees her purpose defeated, and the vital energy flags, the sap of life runs to seed. On the same principle an existence of joyless drudgery seems to drain the springs of health, even at an age when they can draw upon the largest inner resources; hope, too often baffled, at last withdraws her aid; the tongue may be attuned to canting hymns of consolation, but the heart can not be deceived, and with its sinking pulse the strength of life ebbs away. Nine tenths of our city children are literally starving for lack of recreation; not the means of life, but its object, civilization has defrauded them of; they feel a want which bread can only aggravate, for only hunger helps them to forget the misery of *ennui*. Their pallor is the sallow hue of a cellar-plant; they would be healthier if they were happier. I would undertake to cure a sickly child with fun and rye-bread sooner than with tidbits and tedium.

Mirth is a remedy; the remarkable longevity of the French aristocrats,* in spite of their dietetic and other sins, can with certainty be ascribed to the gayety of their pastimes; almost any mode of diversion is better than the deadly monotony of our sabbatarian machine-life; even excursion-trains have added years to the average longevity of our city populations. In a temperature of — 56° Fahr., Elisha Kane kept his men in good health by devoting a part of the long night to burlesques and pantomimes; but, as a sanitary precaution, dramaturgy was only collateral to the substitution of tea for grog; and the most striking illustration of the hygienic effect of merriment is therefore, perhaps,

---

* E. g., Polignac, eighty-one years; Richelieu, eighty-three; Sainte-Pierre, seventy-eight; Chateaubriand, eighty; Lafayette, seventy-eight; Duke of Bassano, eighty-one; Corneille, eighty; Dumouriez, eighty-four; Palinet, eighty-five; Fontenelle, one hundred; Joinville, ninety-one; L'Enclos, eighty-nine; La Maintenon, eighty-four; Rochefoucauld, eighty; Villars, eighty-one; Sully, eighty-one; Montfaucon, eighty-six; Soult, eighty-two; Talleyrand, eighty-four.

the experience of Dr. Brehm, the manager of the Hamburg Zoölogical Garden. Having noticed that the monkeys in the happy-family department generally outlived the solitary prisoners, he concluded to try the Swiss nostalgia-remedy, "fun and cider-punch"; but the liquid stimulants proved superfluous: the introduction of a grapple-swing and a few toys sufficed to reverse the shadow on the dial of death, and man by man the quadrumana recovered from a disease which evidently had been nothing but *ennui*, since the mortuary lists of the last decade showed an almost uniform death-rate throughout the year, except in midsummer, when the monkey-house could be thoroughly ventilated.

Men of a cheerful disposition are generally long-lived, and anything tending to counteract the influence of worry and discontent directly contributes to the preservation of health. Despair can paralyze the energy of the vital functions like a sudden poison, while the fulfillment of a long-cherished hope has effected the cure of many diseases; history abounds with examples of strong men dying of sheer grief,* as well as of a great success giving to others a new lease of life. Even hope can sustain the vital powers under severe trials; the appearance of a distant sail or a leeward coast has often restored the strength of shipwrecked sailors who would have succumbed to another hour of hopeless famine. A mere day-dream of a possible deliverance from toil or captivity prolongs the life of thousands who would not survive an awakening to the realities of their situation.

But "hope deferred" sickens the body as well as

---

* E. g., Isocrates, Kepler, Mehemet Ali, Bajazet, Politianus, Columbus, Maupertuis, Pitt, the two Napoleons, Nicholas I, Joseph II, Platen, Abd-el-Kader, Shamyl, Horace Greeley.

the soul; and, next to the happiness of a life whose labors are their own immediate reward, is the confident anticipation of a period of compensating enjoyments at the end of every day, of every week, and every year, or part of a year. With a few playthings the youngsters of the nursery will find pastimes enough, though even the youngest should have some corner of the house where they can feel quite at home; but the necessity of providing special times and modes of recreation begins with the day when a child is delivered to the taskmaster, when its employment during any considerable part of the twenty-four hours becomes laborious and compulsory. Children under ten should never be kept at school for more than three consecutive hours, unless the variety of the successive lessons forms itself a sort of recreation, as drawing after grammar, or writing alternating with "calisthenics" or vocal exercises. If the principal meal of the day is taken at noon, the midday recess should be extended to at least three hours; otherwise one hour is more than sufficient, especially where the recess sports are diverting enough to forget the school-room for a few minutes. The more completely a special train of thoughts can for a while be dismissed from the mind, with the more profit can it afterward be resumed, for the same reason that the successful practice of any bodily exercise requires a periodical relaxation of the strained muscles. But, if the instinct of rooks and savages can be trusted, the recreation-time, *par excellence*, is the evening hour; and with a little management young and old bondmen of drudgery might consecrate the end of every day to health-restoring sports. All schools ought to close at 4 P. M.; and, till we can enforce the eight-hours labor law, the societies for the prevention of cruelty should liberate at least the younger factory-slaves two hours before the sunset of a summer day, in order to give them a chance for a few minutes' recreation between supper and bed-time. "*Horas non conto, nisi serenas*" was the usual inscription of the Roman sun-dials, but the Arabs of the desert count time by nights instead of days; and for us, too, sunset is the beginning of the most pleasant and most play-inviting hour of the twenty-four; the day's work is done, no fear of interruption damps the merriment of the moment, and to the fatigue of boisterous sports the coming night offers the refuge of rest and sleep. For the same reason the compulsory somnolence of our Quaker-sabbath makes Saturday night the Saturnalia-time of many Christian nations; the Sunday laws have reduced them to amusements which can, and too often ought to, dispense with daylight, and in the larger cities apprentices and factory-boys have the alternative of joining in such night revels or postponing their amusements to the musical resurrection of the saints in light, for the free Saturday is unfortunately confined to primary schools and a few private seminaries. In German schools Saturday is at least a half-holiday; i. e., the scholars are dismissed at noon, and at once make for the fields and woods, except in winter, when the disciples of the Turnerhall assemble on the last afternoon in the week.

With our present helplessness against the lethargic influence of the midsummer heat, the conventional time of the long vacations is well selected, but, if a hoped-for diet and dress reform shall have taught us to pass the dog-days with comfort, it would be more sensible to divide the two months: four free weeks in June, in time for the first huckleberries and butterflies, and four in October—the best season for a long excursion to the paradise of a primitive mountain-range, nowadays about the only sanctuary of Nature where her worshipers can shake their shoulders free from the yoke of prejudice and escape from the atmosphere of hypocrisy to a higher and purer medium. For the children of the poor every city should have a *Kinder-park* —not a ceremonious promenade, with sacred groves and unapproachable grass-plots, but a public play-ground with shade-trees and swings, May-poles, gymnastic contrivances and a free bathing-house, and room for all the free menageries and music-halls which the Peabodies of the future might feel inclined to add. Inactivity is no recreation; we should not spend our leisure hours like machines, whose best relief is a temporary surcease of toil, but like living creatures of the God who intended that the joys of life should outweigh its sorrows. Let us provide healthful pastimes, or the victims of asceticism will resort to vices—dram-drinking, gambling, and secret sins—for even pernicious excitements become attractive as a relief from the insupportable dullness of a canting Quaker life.

Ennui has never made a human being better or more industrious; on the contrary, the hope of a merry evening would inspire a day-laborer with a good-humor and an energy unknown to the languid *resignados* of our present system. The confident expectation even of a physical pleasure imparts to the current of life an onward impulse that seems to react on the mind as well as on every function of the automatic organism; the first Napoleon, who enlivened the tedium of camp-life with Olympic festivities, and did not deem it below his dignity to make his own *maitre de plaisir*, could in return rely on his men to endure fatigues that would have killed the barrack-slaves of his enemies. It is not hard work that drives our young men to seek a Lethe in alcohol: we read of Grecian soldiers marching fifty miles a day in heavy armor; of hunters running down a wild-boar, and of teamsters yoking themselves to a car when their horses had broken down. Many of our New England boys, who go on a whaling cruise rather than die of *ennui*, would gladly consent to work, in the ancient sense of the word, if they could exchange their Pecksniff-day for a Grecian festival. The Aryan nations, too, had their sacred days and sacred rites, but their Nature-worship was the mist that rises from the woods and meadows, and blends with the ethereal hues of the sky; the Hebrew priestcraft dogma is a poison-cloud which for centuries has darkened the light of the sun and blighted the fairest flowers.

In choosing the mode of a child's recreations, it should be borne in mind that their main purpose is to restore the tone of the mind and its harmony with the physical instincts by supplying the chief deficiencies of our ordinary employment. For a hard-working blacksmith, fun, pure and simple, would be a sufficient pastime, while brain-workers need a recreation that combines amusement with physical exercise—the unloosening of the brain-fiber with the tension of the muscles. Emulation and the presence of relatives and school-mates impart to competitive gymnastics a charm which a spirited boy would not exchange for the passive pleasure of witnessing the best circus-performance. Wrestling, lance-throwing, archery, base-ball, and a well-contested foot-race, can awaken the enthusiasm of the Grecian *palæstra*, and professional gymnasts will take the same delight in the equally healthful though less dramatic trials of strength at the horizontal bar. But, on the play-ground, such exercises should be divested from the least appearance of *being a task*—even children can not be happy on compulsion.

There is also too much in-door and in-town work about the present life of our school-boys. Encourage their love of the woods; let us make holidays a synonym of picnic excursions, and enlarge the definition of camp-meetings; of all the known modes of inspiration, forest air and the view of a beautiful landscape are the most inexpensive, especially from a moral stand-point, being never followed by a splenetic reaction. A ramble in the depths of a pathless forest, or on the heights of an Alpenland, between rocks and lonely mountain-meadows, opens well-springs of life unknown to the prisoners of the city tenements.

But the chief curse of our in-door life is, after all, its dullness; and its direct antidote merriment, therefore the chief point about all real recreations. Fun and laughter have become the most effective cordials of our materia medica, and their promotion a most important branch of the science of happiness. There is no such thing as genuine frolic in the stifling atmosphere of a stove-room; the shady lawn in summer and the open hall in winter make a better play-ground than the stuffy nursery; but freedom from restraint is a still more essential element of mirth. Even in the despotic countries of the Old World the representative of the government attends the public *fêtes* in disguise, and, if the schoolmaster wants to watch the recess-sports of his pupils, let him do so unobserved; if you can trust your children at all, trust them not to abuse the freedom of their recreations, or else conduct your surveillance as unobtrusively as possible. Children detest ceremonies; in our etiquette-ridden towns too many boys are aliens under their fathers' roof; give them one hour in the day and one corner in the house where they are really at home, where they can feel that the permission to enjoy themselves is granted as a right rather than as a concession to the foibles of youth. If I had to board my children in an old hull, like Anderson's sea-shell peddler, I would let them store their toy-shells in the caboose, and keep it sacred from the intrusion of the forecastle folk, to let my little ones know that the believers in the divinity of joy, though in a sad minority in this pessimistic world, have rights and perquisites which I mean to maintain against all comers.

It does not cost much to make the little folks happy; time, and permission to use it, is all the most of them ask; but make them sure that the pursuit of happiness is not a contraband affair, but a legitimate and praiseworthy business. Nor can it do any harm to let them accumulate a little stock in trade—marbles, tops, dolls, and magic lanterns, and, if possible, a few pets; in winter-time, and for the bigger boys, a private menagerie of squirrels and gophers is a better aid to domestic habits than a hundred interviews with the home-missionary. Connive at a snow-ball fight or a torn hat; and be sure that a pair of skates, fishing-tackle, and a base-ball outfit are a better investment than a medicine-chest. Make your children happy; all Nature proclaims the plan of a benevolent Creator; let them feel that their life is in harmony with that plan—that existence has a positive value, an attraction that would remain, though the fear of death were removed.

And, above all, let no cloud of superstition darken the sunshine of your Sundays; and, in countries where the knell of the church-bells drives your children from the play-grounds of the city, take them out to the woods and mountains, and let them worship the Cre-

ator in his grandest temple; teach them to love his day **by** making it the happiest day in the week. Or, disregard the bells and brave the consequences: till we can repeal the sabbath **laws, let** us defy them in every way and **at any risk; in** dealing with the despotism of the mythology-mongers, legal obligations are out of the **question; the** right **of** Nature enters the lists against **the right of brutal** force leagued with imposture.

CHAPTER VIII.

REMEDIAL EDUCATION.

"We can not buy health; we must deserve it."—Francis Bichat.

"Prevention is better than cure and far cheaper," said John Locke, two hundred years ago; and the history of medical science has since made it more and more probable that, in a stricter sense of the word, prevention is the only possible cure. By observing the health laws of Nature, a sound constitution can be very easily preserved, but, if a violation of those laws has brought on a disease, all we can do by way of "curing" that disease is to remove the cause; in other words, to *prevent* the continued operation of the predisposing circumstances.

Suppressing the symptoms in any other way means only to change the form of the disease, or to postpone its crisis. Thus, mercurial salves will cleanse the skin by driving the ulcers from the surface to the interior of the body; opiates stop a flux only by paralyzing the bowels—i. e., turning their morbid activity into a morbid inactivity; the symptoms of pneumonia can be suppressed by bleeding the patient till the exhausted system has to postpone the crisis of the disease. This process, the "breaking up of a sickness," in the language of the old-school allopathists, is therefore in reality only an interrupting of it, a temporary interruption of the symptoms. We might as well try to cure the sleepiness of a weary child by pinching its eyelids, or the hunger of a whining dog by compressing his throat.

Drugs are not wholly useless. If my life depended upon a job of work that had to be finished before morning, and the inclination to fall asleep was getting irresistible, I should not hesitate to defy Nature, and keep myself awake with cup after cupful of strong black coffee. If I were afflicted with a sore, spreading rapidly from my temple toward my nose, I should suppress it by the shortest process, even by deliberately producing a larger sore elsewhere, rather than let the smaller one destroy my eyesight. There are also two or three forms of disease which have (thus far) resisted all unmedicinal cures, and can hardly be trusted to the healing powers of Nature—the *lues venerea*, scabies, and prurigo—because, as Claude Bernard suggests, their symptoms are probably due to the agency of microscopic parasites, which oppose to the action of the vital forces a life-energy of their own, or, as Dr. Jennings puts it, "because art has here to interfere—not for the purpose of breaking up diseased action, but for the removal of the cause of that action, the destruction of an active virus that possesses the power of self-perpetuation beyond the dislodging ability of Nature."

But with those rare exceptions it is better to direct our efforts against the cause rather than the symptoms —i. e., in about ninety-nine cases out of a hundred it is not only the safer but also the shorter way to avoid drugs, reform our habits, and, for the rest, let Nature have her course; for, properly speaking, disease itself is a reconstructive process, an expulsive effort, whose interruption compels Nature to do double work; to resume her operations against the ailment after expelling a worse enemy—the drug. If a drugged patient recovers, the true explanation is that his constitution was strong enough to overcome both the disease and the druggist.

Dr. Isaac Jennings,* the greatest pathologist (or, at least, patho-*gnomist*) of our century, was sadly misunderstood, chiefly, I believe, because he called his method the "Let-alone Plan." Prevention Plan, or Unmedicinal Cure, would have been a better word. Diseases do not want to be let alone; they call loudly for relief —not, though, from their own symptoms, which, in fact, are so many alarm-signals, but from the obstacle which has forced the vital process to deviate from its normal course. Pain, in all its forms, is an appeal for help, and the urgency of the appeal corresponds to the degree of the distress; probably, also, to the possibility of relieving that distress. A deadly blow stuns—the vital forces yield without a struggle. The last stage of pulmonary consumption is a comparatively painless *deliquium*—when a conflagration grows uncontrollable, the alarm-bells cease to ring. Yellow-fever doctors give up their patients for lost when the burning headache changes into a lethargic stupor. The last sensations of drowning, strangled, and freezing persons are said to be rather pleasurable than otherwise. In certain cases the appeal for help continues into an apparently hopeless stage of the disease. Apparently, I say: Nature is too practical to waste her efforts on a forlorn hope; her resistance yields to necessity; and, when the art of healing shall devote itself to the exegesis of disease rather than to the exorcism of its symptoms, that rule will probably be found to apply to pathology as well as to chemistry and ethics.

All bodily ailments are more or less urgent appeals

---

* Author of the "Treatise on Medical Reform."

for help; nor can we doubt in what that help should consist. The more fully we understand the nature of any disease, the more clearly we see that the discovery of the cause means the discovery of the cure. Many sicknesses are caused by poisons, foisted upon the system under the name of tonic beverages or remedial drugs; the only cure is to eschew the poison. Others, by habits more or less at variance with the health laws of Nature; to cure such we have to reform our habits. There is nothing accidental, and rarely anything inevitable, about a disease; we can safely assume that nine out of ten complaints have been caused and can be cured by the sufferers (or their nurses) themselves. "God made man upright"; every prostrating malady is a deviation from the state of Nature. The infant, "mewling and puking in its nurse's arms," is an abnormal phenomenon. Infancy should be a period of exceptional health; the young of other creatures are healthier, as well as prettier, purer, and merrier, than the adults, yet the childhood years of the human animal are the years of sorest sickliness; statistics show that among the Caucasian races men of thirty have more hope to reach a good old age than a new-born child has to reach the end of its second year. The reason is this: the health theories of the average Christian man and woman are so egregiously wrong, that only the opposition of their better instincts helps them —against their conscience, as it were—to maintain the struggle for a tolerable existence with anything like success, while the helpless infant has to conform to those theories—with the above results.

"I have long ceased to doubt," says Dr. Schrodt, "that, apart from the effects of wounds, the chances of health or disease are in our own hands; and, if people knew only half the facts pointing that way, they would feel *ashamed* to be sick, or to have sick children."

A vestige of the hygienic insight which in savages appears to be a gift of Nature, would, indeed, almost obviate the necessity of a treatise on the diseases of infancy; nay, wherever people have got rid of four or five of the grossest physiological prejudices, the art of preserving the health of a healthy-born child is even now a sort of intuition with every true mother; but nurses, physicians, and foster-parents, are often called upon to mend the mistakes of their predecessors and to undertake a task whose less intuitive duties may be facilitated by some of the following hints on remedial education:

Shakespeare's "mewling and puking" representative of babyhood was probably overfed. The representative nurse believes in cramming; babies, like prize-pigs, are most admired when they are ready to die with fatty degeneration. The child is coaxed to suckle almost every half-hour, day after day, till habit begets a morbid appetite, analogous to the dyspeptic's stomach distress which no food can relieve till over-repletion brings on a sort of gastric lethargy.

"Many hand-fed infants, weighing about ten pounds, will swallow one and a half quart of cow's milk in one day," says Dr. Page;* "now, considering the needs of a moderately working man to be equal in proportion to size, a man weighing one hundred and fifty pounds should take fifteen times the quantity swallowed by the infant, or twenty-two and a half quarts—a quart for nearly every hour of the day and night!"

---

* "How to feed a Baby to make it healthy and happy," p. 23.

Vomiting, restlessness, and gross fatness, are some of the symptoms of the surfeit-disease, and its proper cure is—not soothing-sirups, but—fasting. Four nursings a day are enough, five more than enough, and the ejection of milk after suckling is a sure sign that the quantity given at each meal should be diminished. A pint of milk a day is about as much as a dyspeptic infant can really digest, and to cram it merely in order to stop its crying is quite mistaking the cause of its restlessness; a half-starved child will not cry, because the languor of insufficient nutrition is a pleasure compared with the gastric torments of the surfeit-disease. Children actually perishing with hunger will utter from time to time a peculiar sharp cry, almost like the call of a hungry nest-bird, but the first mouthful of food makes them relapse into a sort of dreamy silence.

There are nurslings who get at least four times more milk and pap than they can possibly assimilate, and whose digestive organs have to reject the surplus in a way that would make life intolerable to an adult, though most nurses seem to consider retching and "dripping" as a normal phase of infant life.

Drugs only complicate the disorder: many children whose constitution would have resisted the cramming process succumb to opiates, "surfeit-water" and ipecacuanha; but, unless foul dormitories still further aggravate the evil, each night generally undoes the mischief of the day; the child becomes plethoric with fat; Nature has shifted the burden from the vital organs to the tegumental tissues, and in hopes of final relief manages to hold the fort of life against daily and complicated attacks. Relief comes at last when the nursling is weaned and reduced from ten or twelve to three meals a day. The after-effects of medication may retard recovery for a while, but, the main cause being removed, the morbid symptoms disappear in the course of four or five months.

A less frequent but (through gross maltreatment) often more dangerous disease is scrofula, the cachectic degeneration of the humors resulting from the combined influence of unwholesome food and foul air. In the rural districts of our milk and corn-bread States scrofulous children are as rare as white wolves in the tropics; in Northern Europe the disease is now far less prevalent than formerly; and the operatives of our large cities, in spite of their wretched habitations, might avoid it altogether, or at least obviate its more serious consequences, but for the fatuous quackery which so often turns a transient skin-disease into a chronic lung-complaint. In the middle ages, when science was at its lowest ebb and supernaturalism in full tide, the "king's-evil" was considered an almost unavoidable disease, resisting all common remedies and yielding only to the mandate of royalty—the touch of a legitimate king, supplemented by the mandamus of a clerical exorcist. In the fifteenth century from eight to twelve thousand families per year performed long journeys to the English capital; Charles II, in the course of his reign, touched near a hundred thousand persons. The days on which the miracle was to be wrought were solemnly notified by the clergy of all parish churches (Macaulay's "History of England," Chapter XIV). Traveling was expensive in those days, and, scrofula being distinctively a disease of the poor, nine out of ten patients of the royal doctor had probably come afoot, and often from distances which suggest the explanation of the marvelous cures: the pilgrims left the pest-air of their hovels behind, and Nature availed herself of the respite, as she improves a temporary change from city fumes to the woodland air of some rural retreat whose salubriousness is ascribed to the accidental presence of a nauseous sulphur-spring—the one abnormal thing about the place. The king's-evil patients, as well as the exorcists, ascribed the cure to what Dr. Joel Brown called the *charisma basilicon* —the healing touch of the Lord's anointed—in other words, they believed that the cure of a Yorkshire man's disease depended upon the chance of the Yorkshire man's coming in contact with a Londoner who, perhaps ten or twenty years ago, had undergone the rites of a certain ceremony. Imagination probably helped a little, for after the spread of skepticism " perfect cures became much less frequent," as Dr. Brown naïvely remarks. The *charisma basilicon* has now fallen into utter discredit, but our present method is so little of an improvement that the patients of a future century would probably prefer to resume the Whitehall pilgrimages. Instead of ventilating our houses and abolishing our sauerkraut (the long-notorious cachexia of the ill-housed and ill-fed classes having sufficiently indicated the cause of the malady), we suppress the morbid symptoms by sarsaparilla, iodide of potassium, or patent "medicines": only reliable liver-pills and infallible blood-purifiers—in other words, we believe that the cure of a common disease depends upon the accidental or providentially ordained discovery of some mysterious compound. The bottom error is the same as in the king's-evil delusion, and can be easily traced to the radical fallacy of our speculative dogmas; we still regard sin and disease as something normal, aboriginal, and unavoidable, and expect salvation from mysterious, extra-natural remedies, while the truth of the very contrary is becoming more and more evident, namely, that all evil, including moral and physical unsoundness, is due, and generally traceable, to wholly abnormal causes, and (those causes being removed) recovery the effect of the self-acting and self-regulating laws of Nature. The removal of the cause is a remedy which the sufferers from almost any disease might prescribe for themselves, and here especially: fresh air and abstinence from indigestible food, particularly pickles and fat meat. Pork is not the only unwholesome kind of animal food, for Jews are not exempt from scrofula, and were formerly subject to a still worse skin-disease; and, if we had not forgotten the art of interpreting the language of our instincts, we would not overlook the significancy of the circumstance that ninety-nine per cent of all young children detest every kind of fat meat except in the form of taste-deceiving ragouts. Farmer-boys, who have to share the out-door labors of their parents, can eat with *comparative* impunity many things which only the hardiest of their city comrades can digest: pork, greasy and pickled cabbages, fritters, and salt beef. Even young Hottentots could not eat such stuff without being sooner or later the worse for it, whenever the counteracting hardships of a savage life alternate with a period of physical inactivity. But children afflicted with cachectic symptoms should at once be restricted to a wholly vegetable and non-stimulating diet—farinaceous preparations, boiled legumina, and, if possible, ripe, sweet fruit.

The summer diet of a scrofulous child can not be too *frugal*, in the ancient sense of the word, and, where a supply of ripe tree-fruits can be easily obtained, I should think it the best plan to dispense altogether with made dishes—for a while, even with farinaceous dishes. Parents who have no hesitation in cramming their children with salt pork, beer, and sauerkraut, would shudder at the idea of feeding them on fruit alone, yet the happiest of all visitors to the southern Rhineland are probably the patients of a Swabian *Trauben-Kur*, where dyspeptics, etc., are fed almost exclusively—often for days together quite exclusively—on ripe, sweet grapes. Combined with plenty of exercise in the bracing air of a highland region, the efficacy of the grape-cure surpasses all the miracles of the king's touch. It will cure children, "too scrofulous to look out of their eyes," cheaper and quicker than any nostrums, and has the still greater advantage of eliminating instead of suppressing the virus.

Those who deny the pharmaceutic efficacy of the homœopathic sugar-pellets can not deny that, in this case, homœopathy has proved the possibility of curing diseases without any drugs at all—merely by a change of diet and regimen. Frugality, abstinence, bathing, ventilation, cold water, and exercise in the open air, have already superseded half the *materia* of the old medical dogmatists, and personal experience has convinced me that the following diseases of children are amenable to a strictly hygienic treatment.

The vicissitudes necessarily incident to an out-door and primitive mode of life are never the first causes of any disease, though they may sometimes betray its presence. *Bronchitis*, nowadays perhaps the most frequent of all infantile diseases, makes no exception to this rule; a draught of cold air may reveal the latent progress of the disorder, but its cause is long confinement in a vitiated and overheated atmosphere, and its proper remedy ventilation and a mild, phlegm-loosening (saccharine) diet, warm sweet milk, sweet oatmeal-porridge, or honey-water. Select an airy bedroom, and do not be afraid to open the windows; among the children of the Indian tribes who brave in open tents the terrible winters of the Hudson Bay Territory, bronchitis, croup, and diphtheria are wholly unknown; and what we call "taking cold" might often be more correctly described as taking *hot;* glowing stoves, and even open fires, in a night-nursery, greatly aggravate the pernicious effects of an impure atmosphere. The first paroxysm of *croup* can be promptly relieved by very simple remedies: fresh air and a rapid forward-and-backward movement of the arms, combined in urgent cases with the application of a flesh-brush (or piece of flannel) to the neck and the upper part of the chest. Paregoric and poppy-sirup stop the cough by lethargizing the irritability and thus preventing the discharge of the phlegm till its accumulation produces a second and far more dangerous paroxysm. These second attacks of croup (after the administration of palliatives) are generally the fatal ones. When the child is convalescing, let him beware of stimulating food and overheated rooms. Do not give aperient medicines; costiveness, as an after-effect of pleuritic affections, will soon yield to fresh air and a vegetable diet.

Worms.—Intestinal parasites are symptoms rather than a cause of defective digestion, and drastic medicines (calomel, Glauber's salt, etc.) are merely palliatives; even a change of diet may fail to afford permanent relief if the general mode of life favors a costive condition of the bowels. Like maggots, maw-worms seem to thrive only on putrescent substances, on accumulated ingesta in a state of self-decomposition, and disappear as soon as exercise, cold fresh air, and a frugal diet, have re-established the functional vigor of the digestive organs.

Diarrhœa.—An abnormal looseness of the bowels is an effort of Nature to rid the stomach of some irritating substance, and suggests the agency of a dietetic abuse, either in quantity or in **quality.** An excessive quantum even of the healthiest **food will** purge the bowels like a drastic poison, **unless the** alimentary wants—and consequently **the** assimilative abilities of the system— have **been increased** by active exercise. On the hunting-grounds of the upper Alps, an Austrian sportsman can assimilate a quantity of meat which the kitchen artists of **the** best Vienna restaurant could not have foisted upon the stomach of an indolent burgher. Dysentery medicines can be entirely dispensed with if one can get the patient to try the effect of Nature's two specifics—fasting and pedestrian exercise. Combined they will only fail when opiates have produced an inflammatory condition of the bowels, in which case a grape- or water-cure must precede the more radical remedies. The languor of dysentery is always combined with a fretful restlessness, and should not **be mistaken for the exhaustion** that calls for repose and food: **the patient is safe if we can** fatigue him into actual sleepiness, or anything like a genuine appetite; when the digestive organs announce the **need** of nourishment, they can be relied upon to find **ways** and means to retain it.

Constipation.—A slight stringency of the bowels should **never be** interfered with; in summer-time close stools are **consistent with** a good appetite **and general** bodily vigor. **Aperient** medicines provoke **a morbid** activity of the bowels, **followed by a** costiveness that differs from a summer **constipation as** insomnia differs from a transient sleeplessness. **In** England and the United States the use of laxative drugs has repeatedly **become epidemic,** and in its consequences a true national misfortune;* and a sad majority of otherwise intelligent parents are still afflicted with the idea that children have to "take something"—in others words, that their bowels have to be convulsed with poisons—for every trifling complaint. Constipation is often simply a transient lassitude of the system, a functional tardiness caused by fatigue and perspiration, and very apt to cure itself in the course of two or three days, especially at a change from a higher to a lower temperature. After the third day the disorder demands a change of regimen: cold ablutions, lighter bedclothes, in summer-time removal of the bed to the coolest and airiest available locality, and liberal rations of the most digestible food—bran-bread, sweet cold milk, stewed prunes, and fresh fruit in any desired quantity; *faute de mieux*, cold water and sugar, oatmeal-gruel, and diluted molasses. The legumina, in all their combinations, are likewise very efficient bowel regulators, and common pea-soup is a remedial equivalent of Du Barry's expensive "revalenta Arabica" (lentil-powder). For real dyspepia (rarely a chronic disease of youngsters in their teens), there is hardly any help but rough out-door exercise, daily pedestrian exercise or out-door labor, continued for hours in all kinds of weather. The Graham starvation cure might bring relief in the course of time, but, for one person with passive heroism enough to resist the continual cravings of an abnormal appetite, hundreds can muster the requisite resolution for an occasional

---

* "If the bowels become constipated, they are dosed with pills, with black draughts, with brimstone and treacle, and medicines of that class, almost *ad infinitum*. Opening medicines, by constant repetition, lose their effects, and therefore require to be made stronger and stronger, until at length the strongest will scarcely act at all; . . . the patients become dull and listless, require daily doses of physic until they almost live on medicine."—(H. Chavasse, "Advice to a Mother," p. 388.)

active effort, which will gradually but perceptibly restore the vigor of the system. Drugs only change the form of the disease by turning a confirmed surfeit-habit into a still more obstinate and less commutable alcohol-habit; the vile mixtures sold under the name of "tonic" bitters have never benefited anybody but their proprietors and the rum-sellers, to whose army of victims the patent-medicine dispensaries serve as so many recruiting-offices.

Active exercise is also the only remedy for those *secret vices* whose causes are as often misunderstood as their consequences. The pathologists who ascribe precocious prurience to the effects of a stimulating diet seem to overlook the fact that the most continent nations of antiquity, the Scythians and ancient Germans, were as nearly exclusively carnivorous as our Indian hunting-tribes, the apathy of whose sexual instincts has been alleged in explanation of their gradual extinction.* For the same reason the gauchos of the tropical pampas are an unprolific race, while the Russian mujiks and the sluggish boyars of the Danubian principalities are as salacious as the inert (though frugivorous) natives of Southern Italy. Independent of climate and diet, the continence or incontinence of the different nations, or different classes of any nation, bears an unmistakable proportion to the degree of their indolence. Lazy cities and small, thickly populated islands (Lesbos, Paphos, Cythera, Otaheite) have been most conspicuous for the absence of those virtues which the Grecian allegory ascribed to the goddess of the chase. The *menu* prescribed by the founders of the monastic orders was rather ultra-Grahamite in quality and quantity, yet nei-

* Ludwig, "American Aborigines," p. 128.

ther barley-bread nor the frequent fasts to aid the *minutio monachi* could counteract the effects of deficient exercise; if we can believe the publicists of the Reformation, the *chronique scandaleuse* of Lesbos and Capri was far surpassed in the record of some mediæval convents—and not in the flagrant latitude of Italy alone (Robert Burton's "Anatomy of Melancholy," volume of miscellanies, pp. 449-451, quotations, etc.). Nor can we mistake the significance of the circumstance that sexual aberrations in the years of immaturity are almost exclusively the vice of male children, whose potential energies, with the same diet and the same general mode of life, find no adequate vent in an amount of active exercise nearly sufficient for the constitutional wants of the other sex. Moral lectures are sadly ineffectual in such cases, because, as Gotthold Lessing remarks, vicious passions pervert the constitution of the mind as effectually as they subvert that of the body—"the evil powers blindfold the victims of their altars." A frugal diet may subserve the work of reform, but the great specific is competitive gymnastics, the society and example of merry, manly, and adventurous companions. Crank-work gymnastics won't do; enlist the pride of the young Trimalchion, watch him at play, find out his special *forte*, no matter what—running, jumping, or throwing stones—and organize a sodality for the cultivation of that particular accomplishment. Beguile him into heroic efforts, offer prizes and champion-badges: as soon as manful exercises become a pleasure, unmanning indulgences will lose their attractions. The depressing after-effect of sensual excesses, the dreary reaction, is a chief incentive to the repetition of the vicious act, and the success of all reformatory measures depends at first upon the possibility of relieving this depression by healthful diversion, till, in the course of time, regained mental and bodily vigor will help the remedial tendency of Nature to neutralize the morbid inclination.

"*Rickets*" is a sign of general debility, owing to mal-nutrition during the years of rapid growth. The best physic for a rickety child is milk, bran-bread, and fruit; the best physician, the drill-master of the Turner-hall. Rickety children are apt to be precocious, and till their backs are straightened up their books ought to be thrown aside. Knock-knees, bow-legs, "chicken-breasts," and round shoulders are all amenable to treatment, if the cure be begun in time—during the first three years of the teens, of all ages at once the most plastic and the most retentive of deep impressions.

For the cure of young *topers*, *smokers*, and *gluttons* I am persuaded that punishments are only of temporary avail, and homilies of no use whatever. The most glowing eloquence palls before the suasion of a vicious *penchant*. Here, too, the chances of saving the tempted depend upon the possibility of silencing the tempter —by outbidding his offer. Provide healthful diversions; the victims of the poison-habit yield to temptation when the reaction (following upon every morbid excitement) becomes intolerable. Relieve the strain of that reaction by diverting sports; improvise hunting expeditions and mountain-excursions or Olympic games; between exciting diversions and sound sleep the toper will forget his tipple, and every day thus gained will lessen the danger of a relapse.

It can not be denied that *poison-habits* (the opium-habit as well as "alcoholism") are to some degree hereditary. The children of confirmed inebriates should be carefully guarded, not only against objective temptations, but against the promptings of a peculiar disposition which I have found to be a (periodical) characteristic of their mental constitution. They lack that spontaneous gayety which constitutes the almost misfortune-proof happiness of normal children, and, without being positively peevish or melancholy, their spirits seem to be clouded by an apathy which yields only to strong external excitants. But healthful amusements and healthy food rarely fail to restore the tone of the mind, and, even before the age of puberty, the manifestations of a more buoyant temper will prove that the patient has outgrown the hereditary hebetude, and with it the need of artificial stimulation.

Chlorosis, or green-sickness, is a malignant form of that dyspeptic pallor and languor which one half of our city girls owe to their sedentary occupations in ill-ventilated rooms. The complaint is almost unknown in rural districts, and the best cure is a mountain-excursion, afoot or on horseback; the next best a course of "calisthenics," a plentiful and varying vegetable diet, fun, frequent baths, and plenty of sleep. "Tonic" drugs are sure to aggravate the evil. It is only too well known that a fit of nervous depression can be *momentarily* relieved by a cup of strong green tea. The stimulus goads the weary system into a spasm of morbid activity: the vital strength, sorely needed for a reconstructive process (one of whose phases was the nervous depression), has now to be used to repel a pernicious intruder; and this convulsion of the organism, in its effort to rid itself of the narcotic poison, is mistaken for a sign of returning vigor—the patient "feels so much better." But, as soon as the irritant has been eliminated, the vital energy—diminished now by the expulsive effort—has to resume the work of reconstruction under less favorable circumstances; the patient now "feels so much worse"—by just as much as the reaction following upon the morbid excitement has since increased the nervous depression. In the same way precisely a "tonic" medicine operates upon the exhausted organism, and in the same way its effect —a morbid and transient stimulation—is mistaken for a permanent invigoration.

Pulmonary consumption, in its early stages, is perhaps the most curable of all chronic diseases. The records of the dissecting-room prove that in numerous cases lungs, wasted to one half of their normal size, have been healed, and, after a perfect cicatrization of the tuberculous ulcers, have for years performed all the essential functions of the sound organ. Still, the actual waste of tissue is never perfectly repaired, and fragmentary lungs, supplying the undiminished wants of the whole organism, must necessarily do double work, and will be less able to respond to the demands of an abnormal exigency. But the lungs of a young child of consumptive parents are sound, though very sensitive, and, if the climacteric of the first teens has been passed in safety, or without too serious damage, the problem becomes reduced to the work of preservation and invigoration: the all but intact lungs of the healthy child can be more perfectly redeemed than the rudimentary organs of the far-gone consumptive; the phthisical taint can be more entirely eliminated and the respiratory apparatus strengthened to the degree of becoming the most vigorous part of the organism. The poet Goethe, afflicted in his childhood with spitting of blood and other hectic symptoms, thus completely redeemed himself by a judicious system of self-culture. Chateaubriand, a child of consumptive parents, steeled his constitution by traveling and fasting, and reached his eightieth year. By a relapse into imprudent habits, the latent spark, which under such circumstances seems to defy the eliminative efforts of half a century, may at any time be fanned into life-consuming flames, but in ninety-nine out of a hundred cases it will be found that the first improvement followed upon a change from a sedentary to an out-door and active mode of life. Impure air is the original cause of pulmonary consumption ("pulmonary scrofula," as Dr. Haller used to call it), and out-door life the only radical cure. The first symptoms of consumption are not easy to distinguish from those transient affections of the upper air-passages which are undoubtedly due to long confinement in a vitiated atmosphere: hoarseness, and a dry, rasping cough, rapid pulse, and general lassitude. Spitting of blood and pains in the chest are more characteristic symptoms, but the crucial test is the degree in which the respiratory functions are accelerated by any unusual effort. A common catarrh will not prevent a man from running up-stairs or walking up-hill for minutes together, without anything like visible distress; subjected to the same test, a person whose lungs are studded with tubercles will pant like a swimmer after a long dive, and his pulse will rise from an average of 65 to 110 and even 140 beats per minute. Combined with a hectic flush of the face, night-sweats, or general emaciation, shortness of breath leaves no doubt that the person thus affected is in the first stage of pulmonary consumption. If the patient were my son, I should remove the windows of his bedroom, and make him pass his days in the open air—as a cow-boy or berry-gatherer, if he could do no better. In case the disease had reached its *deliquium* period, the stage of violent bowel-complaints, dropsical swellings, and utter prostration, it would be better to let the sufferer die in peace, but, as long as he were able to digest a frugal meal and walk two miles on level ground, I should begin the out-door cure at any time of the year, and stake my own life on the result. I should provide him with clothing enough to defy the vicissitudes of the seasons, and keep him out-doors in all kinds of weather—walking, riding, or sitting; he would be safe: the fresh air would prevent the *progress* of the disease. But *improve* he could not without exercise. Increased exercise is the price of increased vigor. Running and walking steel the leg-sinews. In order to strengthen his wrist-joints a man must handle heavy weights. Almost any bodily exercise—but especially swinging, wood-chopping, carrying weights, and walking up-hill—increases the action of the lungs, and thus gradually their functional vigor. Gymnastics that expand the chest facilitate the action of the respiratory organs, and have the collateral advantage of strengthening the sinews, and invigorating the system in general, by accelerating every function of the vital process. The exponents of the movement-cure give a long list of athletic evolutions, warranted to widen out the chest as infallibly as French-horn practice expands the cheeks. But the trouble with such machine-exercises is that they are almost sure to be discontinued as soon as they have relieved a momentary distress, and, as Dr. Pitcher remarks in his "Memoirs of the Osage Indians," the symptoms of consumption (caused by smoking and confinement in winter quarters) disappear during their annual buffalo-hunt, but reappear upon their return to the indolent life of the wigwam. The problem is to make out-door exercise pleasant enough to be permanently preferable to the *far niente* whose sweets seem especially tempting to consumptives. This purpose accomplished, the steady progress of convalescence is generally insured, for the differences of climate, latitude, and altitude, of age and previous habits, almost disappear before the advantages of an habitual out-door life over the healthiest in-door occupations.

A tubercular diathesis inherited from both parents need not be considered an insuperable obstacle to a successful issue of the cure. The family of my old colleague, Dr. G——, of Namur, adopted a young relative who had lost his parents and his only brother by febrile consumption, and was supposed to be in an advanced stage of the same disease. The Antwerp doctors had given him up, his complaint having reached the stage of night-sweats and hectic chills, and, though by no means resigned to the verdict of the medical tribunal, he had an unfortunate aversion to anything like rough physical exercise. But his uncle, having from personal experience a supreme faith in the efficacy of the open-air cure, set about to study the character of the youngster, and finally hit upon a plan which resulted in the proudest triumph of his professional career. Pierre was neither a sportsman nor much of an amateur naturalist, but he had a fair share of what our phrenologists call "constructiveness"—could whittle out ingenious toys and make useful garden-chairs from cudgels and scraps of old iron. That proved a sufficient base of operations. The doctor had no farm of his own, and the only real estate in the market was a lot of poor old pastures on a sparsely wooded slope of the Ardennes. Of this pasture-land he bought some ten or twelve acres, including a hill-top with a few shade-trees and a fine view toward the valley of the Sambre. At the first opportunity one of Pierre's garden-chairs was sent up to the lookout point, but rain and rough usage soon reduced it to its component elements—scrap-iron and loose cudgels. Pierre volunteered to repair it, and was supplied with such a variety of material and tools that he made two more chairs, and while he was about it also a rustic round-table with a center-hole, corresponding to the diameter of one of the shade-trees. The hill was only two miles from town, and soon became a favorite evening resort of the G—— family; but the road was rather steep, and Mrs. G—— appealed to the ingenuity of her constructive nephew: could he not try and make a winding trail by knocking some of the rocks and bushes out of the way? Pierre tried, and his success, the uncle declared, proved him an intuitive engineer, the peer of Haussmann and Brunel. That new road had so increased the value of the old pasture that it would be worth while to put up a pavilion and make it a regular hill-top resort. The only drawback upon the advantage of its situation was the want of good drinking-water; but there was a sort of a spring in an adjoining pasture on the opposite slope of the ridge: would Pierre make an estimate of the number of bricks requisite to wall it up and keep the cattle from muddling it? The requisition proved an underestimate, but Pierre made up the deficiency by collecting a lot of passably square stones. The water now became drinkable, and somehow the rumor got abroad that Pierre had *discovered* the spring, whereupon his uncle's neighbor urged him to exercise his talent for the benefit of his valley-meadow, in all but the want of water the best pasture in the parish. Pierre selected a spot where a lot of day-laborers were set to work and actually struck water—by digging deep enough. The gratitude of the farmer was almost too demonstrative for the modest lad, who, however, agreed with his uncle that a talent of that sort might make its possessor a public benefactor, and ought to be cultivated. Would Pierre undertake to locate a well on his uncle's hill-pasture, a little nearer to the lookout point? The brick-spring was too far down, and it would be so convenient to have water on one's own premises! Judging from analogies, the young hydrologist fixed upon a spot at the junction of two ravines, but too near the upper boundary of arboreal vegetation, and, after digging down to a stratum of dry sand-stone detritus, the workmen gave up the job in disgust. But Pierre himself would not yield his point, and offered to dig the well alone if they would give him time, and a boy to turn the windlass of the sand-bucket. His wish was granted, and, before he had been a week at work, his asthma had left him, his digestion improved, and his appetite became ravenous. The well-project had finally to be relinquished, but his uncle consoled him by purchasing the adjoining lot and letting him make a winding road from the brick-spring to the hill-top. The road was built, but Pierre indorsed the opinion of a professional engineer that the well-hole, too, would be full of water if the woods of the upper ridge had not been so ruthlessly destroyed, and that the replanting of forest-trees along the line of the subterranean water-courses would not only replenish the springs but redeem the arid pastures of the foot-hills. The doctor controverted that point, but—just for the sake of experiment—procured a hundred beech-tree saplings, which Pierre planted and watered with untiring assiduity. Some sixty per cent of the trees took root, to the unending astonishment of the uncle, who now declared that his confidence in the fertility of the ridge-land had increased to a degree which encouraged him to try his luck with orchard-trees. They procured a lot of young apple, almond, and apricot trees, about two hundred of each, and planted them along the line of the suppositive watercourses. Pierre superintended the work, and was kept so busy for the next eighteen months that he had no time to be sick for a single day. The boy that was given up by the Antwerp doctors is now a well-to-do horticulturist, able to climb without a stop the steepest ridge in the Ardennes and to fell a forty-years oak-tree in twenty minutes!

In the beginning of this chapter I have mentioned two forms of disease which, thus far, have not proved amenable to the hygienic (non-medicinal) mode of treatment, though it has already been ascertained that a mild vegetable demulcent—sarsaparilla, for instance—is as efficacious in those cases as the virulent mercurials of the old school. Antidotes and certain anodynes will, perhaps, also hold their own till we find a way of producing their effects by *mechanical* means. But, with these few exceptions, I will venture the prediction that, before the middle of the twentieth century, the internal use of drugs will be discarded by all intelligent physicians.

"If we reflect upon the obstinate health of animals and savages," says Dr. Schrodt, "upon the rapidity of their recovery from injuries that defy all the mixtures of materia medica; also upon the fact that the homœopathists cure their patients with milk-sugar and mummery, the prayer-Christians with mummery without milk-sugar, and my followers with a milk-diet without sugar or mummery—the conclusion forces itself upon us that the entire system of therapeutics is founded upon an erroneous view of disease."

And, moreover, I believe that the chief error can be

accounted for: it is founded upon our erroneous view of the cause and cure of evil in general. Translated into plain speech, the foundation-principle of our system of ethics is this: that all natural things, especially our natural instincts, are essentially evil, and that salvation depends upon mysterious, anti-natural, and even supernatural remedies. This bottom-error has long biased all our physical and metaphysical theories. The use of our reasoning powers is naturally as agreeable as the exercise of any other normal function: the anti-naturalists declared war against free inquiry, assured us that the study of logic and natural science is highly dangerous, and that the seeker after truth must content himself with the light of ghostly revelations. We have since ascertained that the ghosts are grossly ignorant of all terrestrial concernments, and that their reports on the supramundane state of affairs are, to say the least, suspiciously conflicting.

In all but the vilest creatures the love of freedom is as powerful as the instinct of self-preservation; the anti-naturalists inculcated the dogma of implicit submission to secular and spiritual authorities. The experiment was tried on the grandest scale, and the result has demonstrated that blind faith leads to idiocy, and that absolute monarchs must be absolutely abolished.

The testimony of our noses justifies the opinion that fresh air is preferable to prison-smells; the anti-naturalists informed us that at various seasons of the year, and every night, the out-door atmosphere becomes mortiferous, and that sleepers and invalids ought to be confined in air-tight apartments. We believed, till we found that the most implicit believers got rotten with scrofula.

Animals seem to live and thrive on the principle

that palatable food recommends itself to the stomach, and that repulsive things ought to be avoided. The anti-naturalists reversed the maxim, and assured us that sweetmeats, uncooked vegetables, cold water, drunk when it tastes best—i. e., on a warm day—raw fruit, etc., are the causes of countless diseases, and that the execrable taste of a drug is not the least argument against its salubriousness. During the middle ages parents used to dose their children with brimstone and calomel, "to purify their blood," and, for the same purpose, the most nauseous mineral springs of every country are still pumped and bottled for the benefit of invalids. There is not a poison known to chemistry or botany but has been, and is still, daily prescribed as a health-giving substance, and, in the form of pills, drops, or powders, foisted upon a host of help-seeking invalids. But, since the revival of free inquiry, we have compared the statements of ancient historians and modern travelers, and it appears that the healthiest nations on earth have preserved their health on the principle that guides our dumb fellow-creatures, and would guide our children if they were permitted to follow their inclinations. An overwhelming testimony of facts has proved that the diseases of the human race can be cured easier without poison-drugs—easier in the very degree that would suggest the suspicion that every ounce of poison ever swallowed for remedial purposes has increased the weight of human misery.* And that same suspicion is

---

* "It is unnecessary for my present purpose to give a particular account of the results of homœopathy; . . . what I now claim with respect to it is, that a wise and beneficent Providence is using it to expose and break up a deep delusion. In the results of homœopathic practice, we have evidence in amount, and of a character sufficient, most incontestably to establish the fact that disease is a restorative operation, or renovating

forced upon us by very cogent *a priori* reasons. If the testimony of our senses helps us to select our proper food, and warns us against injurious substances, have we any reason to suppose that such salutary intuitions forsake us at the time of the greatest need—in the hour of our struggle with a life-endangering disease? Shall we believe that at such times our sense of taste *warns us* against salubrious substances? And does it not urgently warn us against ninety-nine out of a hundred "medicines"? Shall the sick believe that an all-wise Creator has staked the chances of their recovery upon the accident of their acquaintance with Dr. Quack's Quinine Bitters or Puff & Co.'s Purgative Pills? Yet, is it possible to mistake the analogy between the remedial theories of our nostrum-mongers and the alleged moral "plan of salvation"? Is not the key-note of the Semitic dogma mistrust of our natural instincts and reliance upon abnormal remedies—mummeries, mysteries, and miracles?

Poison-mongers, physical or spiritual, will cease to be in request whenever their customers begin to suspect that this world of ours is governed by laws, and not by special acts of intervention; that sickness can be cured only by conformity to those laws, and not by drugs and

process, and that medicine has deceived us. The evidence is full and complete. It does not consist merely of a few isolated cases, whose recovery might be attributed to fortuitous circumstances, but it is a chain of testimony fortified by every possible circumstance. . . . All kinds and grades of disease have passed under the ordeal, and all classes and characters of persons have been concerned in the experiment as patients or witnesses; . . . *while the process of infinitesimally attenuating the drugs used was carried to such a ridiculous extent that no one will, on sober reflection, attribute any portion of the cure to the medicine.* I claim, then, that homœopathy may be regarded as a providential sealing of the fate of old medical views and practice."—(Isaac **Jennings,** M. D., " Medical Reform," p. 247.)

prayers—i. e., anti-natural and supernatural remedies. To the children of Nature all good things are attractive, all evil repulsive: the laws of God proclaim and avenge themselves; the Author of this logically-ordered universe can never have intended that our salvation should depend upon the accident of our acquaintance with the dogmas of an isolated act of revelation; and, as surely as the germ of the hidden seed-corn finds its way through night to light, the unaided instincts of the lowliest islander would guide him safely on the path of moral and physical welfare.

These words would be truisms if Truth had not been a contraband for the last eighteen hundred years: To nine tenths of our Christian contemporaries God's most authentic revelation is still a sealed book; and, before any reformer can hope to turn this chaos of vice, superstition, and quackery, into anything like a cosmos, he *must* convince his fellow-men that the study of Nature has to supersede the worship of miracles, even though that conviction should imply that the fundamental dogmas of our priest-religion are perniciously false.

CHAPTER IX.

HYGIENIC PRECAUTIONS.

"Dangers we can not avoid we must learn to defy."—LESSING.

CREATURES in a state of nature can almost dispense with sanitary precautions; Providence has secured their safety in that respect. Animals are born with the instinct that enables them to distinguish wholesome from injurious plants. In the wilderness, where the neighborhood of man does not tempt them to brave the winter of the higher latitudes, most birds emigrate in time to avoid its rigors; those that stay can rely on their feather-coats; natural selection has adapted their utmost power of endurance to the possible extremes of the atmospheric vicissitudes. The sexual instinct of wild animals is limited to certain seasons and months that preclude the possibility of their young being born at any but the most favorable time of the year. From birth to death the children of Nature can trust themselves to the guidance of their hereditary inclinations; all the contingencies of their simple lives have been amply provided for.

These provisions do not apply exclusively to a state of affairs which the agency of man has in so many ways modified or even reversed; still, it would seem as if Nature had failed to make adequate allowance for the possibility of certain perils incident to our artificial mode of life. This fact is perhaps most strikingly illustrated by the treacherous *non-repulsiveness* of certain mineral poisons. The offensive taste of poisonous plants seems to be proportioned to the degree of their noxiousness; hemlock, strychnine, and opium are forbiddingly nauseous, even in the smallest quantities. A drop of prussic acid fills a whole room with its bitter aroma. But arsenious acid is tasteless and odorless, and so unsuspicious to the most wary animals that its name has become a synonym of ratsbane. The reason is apparently this: that Providence (*or* "natural selection") has endowed animals with a protective antipathy against all poisons they could possibly mistake for comestibles, but not against such *out-of-the-way* things as arsenic or sugar of lead, nor against the mixtures by which the art of man has disguised the taste of naturally unpalatable substances. Coffee, without sugar and milk, "straight and strong," as the Turks drink it, would hardly tempt a Christian school-boy; mixed, it can be made seductive enough to deceive even the *ex-officio* opponents of the stimulant-habit. In such commixtures as milk-punch, beer-soup, "Scutari sherbet," the taste—though not the effect—of alcohol almost disappears; the Algeria trappers catch monkeys with a *mélange* of rum and manna-sirup. A famous cook of the "Frères Provençeaux" used to boast his ability of compounding delightful ragouts from meat in *any* state of decomposition. Early habits and the influence of evil examples also tend to corrupt the integrity of that physical conscience whose arbitrations form the health-code of our dumb fellow-creatures. In large cities the panders of vice vie in the art of making their poisons attractive, and, where such dangers can not be avoided, it is always the safest plan to meet and master them in time.

Early impressions are very enduring, and can make

useful habits as well as evil ones a sort of second nature. In order to forestall the chief danger of in-door life, make your children love-sick after fresh air; make them associate the idea of fusty rooms with prison-life, punishment, and sickness. Open a window whenever they complain of headache or nausea; promise them a woodland excursion as a reward of exceptionally good behavior. Save your best sweetmeats for out-door festivals. By the witchery of associated ideas a boy can come to regard the lonely shade-tree as a primary requisite to the enjoyment of a good story-book. "*Or, mes pensées ne voulent jamais aller qu'avec mes jambes,*" says Rousseau ("Only the movement of my feet seems to set my brains a-going"), and it is just as easy to think, debate, rehearse, etc., walking as sitting; the peripatetic philosophers derived their name from their pedestrian proclivities, and the Stoic sect from their master's predilection for an open porch. Children who have been brought up in hygienic homes not rarely "feel as if they were going to be choked" in unventilated rooms, and I would take good care not to cure them of such salutary idiosyncrasies.

Every observant teacher must have noticed the innate hardiness of young boys, their unaffected indifference to wind and weather. They seem to take a delight in braving the extremes of temperature, and, by simply indulging this *penchant* of theirs, children can be made weather-proof to an almost unlimited degree; and in nothing else can they be more safely trusted to the guidance of their protective instincts. Don't be afraid that an active boy will hurt himself by voluntary exposure, unless his chances for out-door play are so rare as to tempt him to abuse the first opportunity. Weather-proof people are almost sickness-proof; a merry hunting-excursion to the snow-clad highlands will rarely fail to counteract the consequences of repeated surfeits; even girls who have learned to brave the winter storms of our Northwestern prairies will afterward laugh at "draughts" and "raw March winds." Winter is the season of lung-affections, the larger part of them induced by long confinement in a vitiated atmosphere; the part caused by light winter clothes is smaller than most people imagine. I have weathered a good many winters without fur caps and woolen shawls, and I ascribe my immunity to the circumstance that my guardian made it a rule never to force us to wear such things. The Moslems rarely eat before they have washed their hands, and a rather unscrupulous frontier Turk assured me that in his case the practice had nothing to do with superstition; it had become a physiological habit, whose omission, he had found, would produce a fit of very realistic nausea. In the same way more comprehensive ablutions may become a physiological necessity: there are people who owe their sound sleep and other sound things to their inability to go to bed without a sponge-bath. The habit can be formed in one summer.

The dietetic instincts of a rationally educated person should obviate the necessity of special precautions, but in large cities, where temptations walk in disguise, the welfare of inexperienced children may require additional safeguards. In the first chapter of this series I have enumerated the chief arguments of the vegetarian school. Among the incidental advantages of their system it might be mentioned that a purely vegetable diet is the most effectual precaution against a danger which only in one of its exceptional forms was lately brought home to us by the trichina panic. Flesh-eaters always run a risk of inoculating themselves with the germs of the various diseases which both beef- and man-flesh is heir to, consumption especially, and several disorders arising from the corruption of the humors, by the use of decayed or fermented food. Sausage-makers, like trance-mediums, never divulge their trade-secrets, but it is a suggestive fact that, in the Anglo-German cities of this continent, the scrofulous and decrepit old females of the bovine race are known by the name of Bologna cows. Abstinence from *Wurst*, boarding-house hash, and mince-pies, may diminish the danger, but abstinence from all animal food is the safer plan and the easier one. If children were restricted to a vegetable or semi-animal diet (milk, eggs, etc.), I doubt if many of them would afterward choose to overcome that instinctive repugnance to flesh-food expressed in the original meaning of the word *frugality*. The Romans of the Cincinnatian era, though entirely free from Buddhistic scruples, seem to have eschewed animal food for sanitary reasons. Children with a phthisical taint are certainly better off without it. Give them eggs and all the available vegetable fat they can digest, but no flesh nor milk of anyways doubtful origin. Two or three families of moderate means might rent a bit of pasture-land, and divide the milk of a healthy country cow. The sanitary condition of a single animal could be ascertained by any competent farrier, but the control of a wholesale meat-market will always be more or less perfunctory.

Principiis obsta is probably the wisest maxim ever expressed in two words, and I believe that the poison-problem will be ultimately solved on that principle. The work of reform must begin in the nursery; and, under circumstances where we can not keep temptations from our door, we must make our children temptation-proof, inspire them with an indelible abhorrence of drunkenness and poison-slavery of every kind.

"I still find the Laconic method the shortest," writes a friend of mine, alluding to the Spartan plan of warning boys by the example of a drunken Helot. He used to interest his boy in the *modus operandi* of alcohol, opium, etc., and then take him out, and, under some pretext or other, drop into a slum-saloon on Saturday night, or a police-court on Monday morning, to give him a practical illustration of his theory. Whenever they saw the poison displayed in an attractive form, on ornamental sign-boards or in the gorgeous bottles of druggists and hotel-keepers, they would study the well-baited trap with a peculiar interest, and go their way rejoicing, as in the possession of an invaluable secret. The result was that the boy became "aggressively virtuous," and used to button-hole visitors in order to lecture them on the causes and consequences of the popular delusion.

Even city boys do not often contract the nicotine habit till after their twelfth year, and a fit of tobacco-nausea before that time generally induces a forbidding reaction not easy to outgrow. I remember the case of a brutal tavern-keeper who tried to accustom his son to the fumes of Alsatian leaf-tobacco (*vulgo Stinkewitz*), and the unexpected result of his last experiment. He took the lad on a stage-coach trip from Colmar to Metz, and induced the postillion to take in a few extra passengers, whom he treated to clay pipes and Stinkewitz. He then closed the windows, and in less than twenty minutes his son turned deadly pale, and would have fainted if he had not found relief in a violent fit of retching. If he had loathed Stinkewitz before, he now dreaded it, and six years after, when he was apprenticed to a tanner, he surprised his master by asking, as for a special favor, that they would not force him to smoke leaf-tobacco. Frederick the Great, too, ascribed his abhorrence of the weed to the choking tobacco-fumes of the Wusterhauser club-room, where the boon companions of his awful parent used to indulge from 5 to 12 P. M. It is not necessary to suffocate a child with nicotine-fumes, but it can do no harm to take him once in a while to a smoker's den, to sniff the "pestilent and penal fires," and let him glory in his blest exemption.

Coffee and tea temptations, pungent spices, etc., may be forestalled in the same way; much is gained if the dietetic innocence of a child has been preserved to the end of the fourteenth year, the age when routine habits first become physiologically confirmed. The habits of the last years of growth become ingrained, as it were, with the constitution of the body, and will bias the physical inclinations of all after-years; circumstances may oblige a man to conform to the customs of a foreign country, the rules of a regimental mess, etc., but, upon the first opportunity of regulating his own regimen, the habits of his boyhood will reassert themselves, even in regard to the time and number of his daily meals. I know from personal experience the unspeakable advantage of having a constitutional predilection for postponing the principal meal till the day's work is done. It was the plan of the ancient Greeks, and to their followers every day is its own reward—the symposium, and the long, undisturbed *siesta* a daily festival. It almost doubles a man's working capacity, by saving him the dire daily struggle between duty and the after-dinner drowsiness. Children who have tried the two methods will rarely hesitate in their choice.

Give them a lunch at twelve o'clock, and for breakfast a crust of sweet bran-bread, the coarser the better. A hard crust is the best possible dentifrice. I never could get myself to believe in the natural necessity of a tooth-brush. The African nations, the Hindoos, the natives of Southern Europe, the South-Sea Islanders, the Arabs, the South American vegetarians, in short, three fourths of our fellow-men, besides our next relatives, the frugivorous animals, have splendid teeth without sozodont. I really believe that ours decay from sheer disuse; the boarding-house *homo* lives chiefly on pap—wants all his meats soft-boiled, and growls at cold biscuit or an underdone potato; in other words, he delegates to the cook the proper functions of his teeth. We hear occasionally of old men getting a second, or rather third, set of teeth. I met one of them in Northern Guatemala, and ascertained that he had become toothless during a twelve years' sojourn in a sea-port town, and that he got his new set upon his return to his native village, where circumstances obliged him to resume the hard corn-cake diet of his boyhood years. His teeth had reappeared, as soon as their services were called for, and would probably never have absented themselves if a pap-diet had not made them superfluous. An artificial dentifrice will certainly keep the teeth white, but that does not prevent their premature decay; disuse gradually softens their substance, till one fine day the hash-eater snaps his best incisor upon an unexpected piece of bone. Every old dentist knows hundreds of city customers whom the daily use of a tooth-brush did not save from the necessity of applying, before the end of the fortieth year, for a complete "celluloid set." I do not say that a soft tooth-brush and such dentifrices as oatmeal or burned arrow-root can do any harm, but, for sanitary purposes, such precautions must be supplemented by *dental exercise*. Let a child invigorate its teeth by chewing a hard crust, or, better yet, a handful of "St. John's bread" or carob-beans, the edible pod of the *Mimosa siliqua*. Children and whole tribes of the northern races seem to feel an instinctive desire to exercise their teeth upon some solid substance, as pet squirrels will gnaw the furniture if you give them nut-kernels instead of nuts. Thus Kohl tells us that the natives of Southern Russia are addicted to the practice of chewing a vegetable product which he at first supposed to be pumpkin or melon seeds, but found to be the much harder seed of the Turkish sunflower (*Helianthus perennis*). Their national diet consists of milk, *kukuruz* (hominy, with butter, etc.), and boiled mutton, and they seem to feel that their Turkoman jaws need something more substantial. The school-boy habit of gnawing pen-holders, finger-nails, etc., may have a similar significance. The *Mimosa siliqua* would yield abundantly in our Southern States, and its sweet pods would make an excellent substitute for chewing-gum. Our practice of sipping ice-cold and steaming-hot drinks, turn about, has also a very injurious effect upon the brittle substance that forms the enamel of our teeth; no porcelain-glaze would stand such abuse for any length of time, and experience has taught hunters and dog-fanciers that it destroys even the bone-crushing fangs of the animal from which our canine teeth derive their name.

Various diseases of the eye, including myopia, strumous and catarrhal ophthalmia, are due to a scrofulous diathesis, and sometimes to a general debility, and can be radically cured only by out-door exercise and a more nutritious diet. But a transient " weak-sightedness "

(*Schwach-sichtigkeit*, as the Germans call it) is eminently a disease of the school-room, caused by a persistent abuse of the eyes, poring for hours together over a spelling-book or writing by the light of a flickering candle (much worse than twilight), as well as by the wretched print of our modern dictionaries and cheap cyclopædias. It should be kept in mind that reading and writing, even under the most favorable circumstances, require an effort to which the eye can only very gradually accustom itself. Hereditary influences and the preliminary exercises of the infant's eye, as, in examining picture-books, the first graphic essays with a slate-pencil, etc., may help to smooth the difficulty; for it is a fact, attested by the experience of all school-teaching missionaries, that the eyes of an adult, sharp-sighted savage begin to smart and water at the first attempt to decipher the hieroglyphics of his primer. The rudiments ought to be taught in half-hour lessons, with liberal intervals of rest and out-door play; and scrofulous children should never be sent to a public school till after a novitiate of at least six months of home studies. Instruct them never to pore over a book, but to keep the head erect, and, at the first symptoms of dim-sightedness, to let the eyes rest upon some distant object, till the optic nerve has recovered from the short-range strain. The hues of the forest have a wonderfully strengthening influence upon weak eyes, almost like its air upon weak lungs; a woodland excursion is like a return to our native element, the birth-land to whose life-conditions the organs of our ancestors were originally adapted.

Accidents can not be avoided by keeping a boy in his nurse's arms or in a padded family coach. Sooner or later he will have to rely on his own limbs, and it is best that time should find him well prepared. Let him rough it, barefoot and bareheaded; let him climb hills and take short cuts over fences and ravines; every fall, every skinned elbow and bumped head, will impart a lesson in the art of locomotion. Without apprentice-fees of that sort he will never get to be a master. I would even connive at an occasional rough-and-tumble fight with a wild comrade; it will acquaint him with what Talleyrand used to call the "esoteric reason for preserving the peace." Constructiveness, too, often the redeeming propensity of a young scape-grace, has its dangers which had better be mastered than avoided. Instead of lecturing a lad or taking away his pocket-knife for cutting his finger, engage a carpenter to teach him the proper use of edge-tools. Let him have a little workshop of his own, with a lot of scrap-tin, boards, nails, and a five-dollar tool-box. Ten to one that those five dollars will save ten cents a week for dime-novels, and, by-and-by, ten dollars a month for beer and tobacco. If your son should manifest symptoms of the collecting-mania, try to direct it to objects of natural history—herbs, beetles, or butterflies. It may lead to deeper studies, and the love of nature in general. A passion for the study of natural history has often turned the scales in a choice between a farm and a dry-goods prison.

"On a visit to Paris," says Carl Weber ("Democritos," vol. ix, p. 166), "the Mentor of a young man, after a trip to the Jardin des Plantes, should not fail to take him to Bertrand Rival's Anatomical Wax-work Museum. It is no misnomer if Bertrand calls his collection '*Musée physiologique, historique et morale*'—intended not only to instruct but to warn the visitor. *Salus tota illa sapere est.*" As a last resort, perhaps, but hardly before the twentieth year. Precocious prurience is due to

HYGIENIC PRECAUTIONS. 237

causes which can generally be avoided. If you can educate the younger children at home and select their playmates, there is no real danger before the eleventh year of a boy and the ninth of a girl. After that, the following precautions will suffice in all but the unluckiest cases: Let your children have plenty of out-door play, especially in the evening. Wait till they are really sleepy before you send them to bed. Let every child have its own bed, or at least its own bedclothes. Keep your small boys out of the servants' room, and your girls after their tenth year; with girls under ten there is less danger: they are quite sure to tell about any improper thing they see or hear, and the servants seem to know that instinctively. Do not leave them alone with elder children—not even with their own neighbors' and relatives'—till you have satisfied yourself about the character of their new friends. No need of a phrenologist to settle that point: the indications of a child's propensities are not confined to the cranium. Vary the child's diet with the season; put the flesh-pots aside when the approach of the summer solstice threatens the land with the temperatures and temptations of Southern Italy. Let them avoid all greasy-made dishes when it is too warm to take much out-door exercise. And, if possible, cultivate their literary taste to the degree that enables them to appreciate the wit or the common-sense of an author, as well as his imagination, and consequently to loathe unmitigated absurdities. That alone will be an effectual safeguard against ninety-nine dime-novels out of a hundred.

In conclusion, I will add a short miscellany of hygienic rules and aphorisms.

The first thing a child should learn is, to ask for a drink of water. I have seen hand-fed children scream and fidget for hours together, as if troubled by some unsatisfied want, but at the same time rejecting the milk-bottle and pap-dish with growing impatience. In nine such cases out of ten the nurse will either resort to paregoric or try the effect of a lullaby. I need not say that the poison-expedient would be wrong under all circumstances, but, before you try anything else, offer the child a cup of cold water. To a young nursling the mother's breast supplies both food and drink, but farinaceous paps require a better diluent than milk.

If I should name the greatest danger of childhood, I would unhesitatingly say, Medicine. A drastic drug as a remedial agent is Beelzebub in the *rôle* of an exorcist.

Our nursery system, after all reforms, is still far from being the right one—how far, we may infer from the fact that we have not yet learned to make our babies behave as well as young animals.

Tight-swaddling, strait-jacket gowns, and trailing petticoats—restraint, in short, makes our infants so peevish. If we would give them a chance to use their limbs they would have no time to scream.

It would prevent innumerable diseases if people would learn to distinguish a morbid appetency from a healthy appetite. One diagnostic rule is this, that the gratification of the latter is not followed by repentance; another, that the former has to be artificially and painfully acquired: our better nature resists the incipience of a morbid "second nature." After acquitting Nature from all responsibility for such factitious appetites, it may be justly said that a man can find a road to health and happiness by simply following his instincts.

The supposed danger of cold drinks on a hot day is a very expensive superstition. It deprives thousands of people of the most pleasurable sensation the human palate is capable of. It is worth a two hours' *anabasis* in the dog-days to drink your fill at the coldest rock-spring of the mountains.

Bathing in flannel!—I would as soon take ice-cream in capsules. The price of the flannel suit would buy you a season-ticket to a lonely beach.

A disposition to excessive perspiration is often due to general debility, but there is a specific remedy for it. Fill your knapsack with substantials and take a pedestrian trip in midsummer, up-hill, if possible, and without loitering under the shade-trees; in short, give your body something worth perspiring for. After that it will be less lavish of gratuitous performances of that sort. The soldiers of the Legion Étrangère are mostly northmen—Poles, Belgians, and Russians—but upon their return from a year's service in Algiers it takes a long double-quick under a Mediterranean sun to drill them into a sweat.

"A catarrh is the beginning of a lung-disease." It would be the end of it if we did not aggravate it with nostrums and fusty sick-rooms.

Somehow or other we must have abused our teeth shamefully before Nature had to resort to such a veto as toothache.

A tooth pulled in time saves nine.

"If you doubt whether a contemplated act is right or wrong," says Zoroaster, "it is the safest plan to omit it." Let dyspeptics remember that when they hesitate at the brink of another plateful.

The digestion of superfluous food almost monopolizes the vital energy; hence the mental and physical indolence of great eaters. Strong-headed business-men manage to conquer that indolence, but only by an effort that would have made the fortune of a temperate eater.

A glutton will find it easier to reduce the number of his meals than the number of his dishes.

Highland children are the healthiest, and, even starving, the happiest. "There is no joy the town can give like those it takes away."

Paracelsus informs us that the composition of his "triple panacea" can be described only in the language of alchemistic adepts. Nature's triple panacea is less indescribable—fasting, fresh air, and exercise.

A banquet without fruit is a garden without flowers.

The best stuff for summer-wear: one stratum of the lightest mosquito-proof linen.

"Do animals ever go to the gymnasium?" asks an opponent of the movement-cure. Never: they have no time—they are too busy practicing gymnastics out-doors.

Descent from a long-lived race is not always a guarantee of longevity. A far more important point is the sanitary condition of the parents at the birth of the child. Pluck, however, is hereditary, and has certainly a prophylactic, a "health-compelling" influence.

The first gray hairs are generally a sign of *dear-bought* wisdom.

The "breaking-up" of a pulmonary disease could often be accomplished by breaking the bedroom-windows.

Death, formerly the end of health, is nowadays the end of a disease.

Dying a natural death is one of the lost arts.

There seems to be a strange *fatum* in the association of astronomy with humbug: formerly in horoscopes, and now in patent-medicine almanacs.

A patent-medicine man is generally the patentee of a device for selling whisky under a new name.

A "chronic disease," properly speaking, is nothing but Nature's protest against a chronic provocation. To say that chronic complaints end only with death, means, in fact, that there is generally no other cure for our vices.

Every night labors to undo the physiological mischief of the preceding day—at what expense, gluttons may compute if they compare the golden dreams of their childhood with the leaden torpor-slumbers of their pork and lager-beer years.

If it were not for calorific food and superfluous garments, midsummer would be the most pleasant time of the year.

CHAPTER X.

POPULAR FALLACIES.

"A national superstition is a national misfortune. No pious fraud has ever advantaged the world, for every popular delusion becomes the mother of a noxious and numerous progeny."—HELVETIUS.

LOGICIANS distinguish between inferential and presumptive fallacies, the first being founded upon illogical conclusions from correct premises, the second upon logical conclusions from incorrect premises. With few exceptions the most mischievous popular delusions of all ages have arisen from the latter—the "presumptive" fallacies. Where their own interests are involved, men seem gifted with an instinctive faculty for looking through the tricks by which a word-juggler appears to support his sophisms with axioms known to be true, but, where that knowledge itself has been falsified (by repeating fictions till they assume the semblance of truisms), all thus biased will accept as sound whatever logical superstructure dupes or impostors may choose to erect upon such sham facts. If a man had been persuaded that cold is a panacea, he would naturally conclude that Siberia must be the healthiest country in the world. In Hindostan, where the sanctity of horned cattle is an established dogma, no true believer would hesitate to indict an irreverent bull-driver for blasphemy, or to preserve a beefsteak as a sacred relic. As long as the Bible passed for infallible, it seemed perfectly logical to ascribe diseases to witchcraft and their cure to prayer, to regard a man's natural instincts as his natural foes, to deny the difference between one and three, and treat mathematicians as enemies of the human race. The systematic application of spurious principles has led to strange results, and latterly to still stranger disputes concerning the propriety of acknowledging the failure, and the best way of compromising the consequences; but such controversies could often be simplified by tracing the effects to their causes. Ill-founded buildings are naturally shaky. Still, people dislike to be lectured on the chronic dilapidation of their parlor-walls. But he who succeeds in exposing the rottenness of the foundation-timbers will need no specious arguments to demonstrate the expediency of removing the household goods to a safer place.

For many centuries the training of the young was almost monopolized by the propagandists of that most terrible of all delusions, the natural-depravity dogma, and our whole system of practical education is still interwoven with the following fallacies, all more or less deeply-rooted upshots of that dogma:

1. THE LEADING-STRINGS FALLACY.—From the moment a child is born, he is treated on the principle that all his instincts are essentially wrong, that Nature must be thwarted and counteracted in every possible way. He is strapped up in a contrivance that he would be glad to exchange for a strait-jacket, kept for hours in a position that prevents him from moving any limb of his body. His first attempts at locomotion are checked; he is put in leading-strings, he is carefully guarded from the out-door world, from the air that would invigorate his lungs, from the sports that would develop his muscles. Hence, the peevishness, awkwardness,

and sickliness of our young aristocrats. Poor people have no time to imitate the absurdities of their wealthy neighbors, and their children profit by what the model nurse would undoubtedly call neglect. Indian babies are still better off. They are fed on bull-beef, and kicked around like young dogs; but they are not swaddled, they are not cradled, and not dosed with paregoric; they crawl around naked, and soon learn to keep out of the way; they are happy, they never cry. If we would treat our youngsters in the same way, only substituting kisses and bread for kicks and beef, they would be as happy as kids in a clover-field, and moreover they would afterward be hardier and stronger. Every week the newspapers tell us about ladies tumbling down-stairs and breaking both arms; boys falling from a fence and fracturing their collar-bones. From what height would a young Comanche have to fall to break such bones—not to mention South-Sea Island children and young monkeys? The bones of an infant are plastic: letting it tumble and roll about would harden the bony tissue; guarding it like a piece of brittle crockery makes its limbs as fragile as glass. Christian mothers reproach themselves with neglecting their duty to their children if they do not constantly interfere with their movements, but they forget that in points of physical education Nature herself is such an excellent teacher that the apparent neglect is really a transfer of the pupil to a more efficient school.

2. THE NOSTRUM FALLACY.—When a child complains of headache, lassitude, or want of appetite, the nurse concludes that he must "take something." If the complexion of a young lady grows every day paler and pastier, her mother will insist that she must "get something" to purify her blood. If the baby squeals

day and night, a doctor is sent for, and is expected to "prescribe something." What that something should be, the parents would be unable to define, but they have a vague idea that it should come from the drug-store, and that it can not be good for much unless it is bitter or nauseous. Traced to its principles their theory would be about this: "Sickness and depravity are the normal condition of our nature; salvation can come only through abnormal agencies; and a remedy, in order to be effective, should be as anti-natural as possible." Perfectly logical from a Scriptural point of view. But Nature still persists in following her own laws. Her physiological laws she announces by means of the instincts which man shares with the humblest of his fellow-creatures, and health is her free gift to all who trust themselves to the guidance of those instincts. Health is not lost by accident, nor can it be repurchased at the drug-store. It is lost by physiological sins, and can be regained only by sinning no more. Disease is Nature's protest against a gross violation of her laws. Suppressing the symptoms of a disease with drugs means to silence that protest instead of removing the cause. We might as well try to extinguish a fire by silencing the fire-bells; the alarm will soon be sounded from another quarter, though the first bells may not ring again till the belfry breaks down in a general conflagration. For the laws of health, though liberal enough to be apparently plastic, are in reality as inexorable as time and gravitation. We can not bully Nature, we can not defy her resentment by a fresh provocation. Drugs may change the form of the disease—i. e., modify the terms of the protest—but the law can not be baffled by complicating the offense: before the drugged patient can recover, he has to expiate a double sin—the medicine and the original cause of the disease. But shall parents look on and let a sick child ask in vain for help? By no means. Something is certainly wrong, and has to be righted. The disease itself is a cry for help. But not for drugs. Instead of "*taking* something," something ought to be *done*, and oftener something habitually done ought to be *omitted*. If the baby's stomach has been tormented with ten nursings a day, omit six of them; omit tea and coffee from the young lady's *menu;* stop the dyspeptic's meat-rations, and the youngster's grammar-lessons after dinner. But open the bedroom-windows, open the door and let your children take a romp in the garden, or on the street, even on a snow-covered street. Let them spend their Sundays with an uncle who has a good orchard; or, send for a barrel of apples. Send for the carpenter, and let him turn the nursery or the wood-shed into a gymnasium. In case you have nothing but your bedroom and kitchen, there will still be room for a grapple-swing; the Boston Hygienic Institute has patented a kind that can be fastened without visible damage to the ceiling. If the baby won't stop crying, something ought to be done about it. Yes, and as soon as possible: remove the strait-jacket apparatus, swaddling-clothes, petticoat, and all, spread a couple of rugs in a comfortable corner, and give the poor little martyr a chance to move his cramped limbs; let him roll, tumble, and kick to his heart's content, and complete his happiness by throwing the paregoric-bottle out of the window.

3. THE STIMULANT FALLACY.—Eight hours of healthy sleep are sufficient to restore the energy expended in an ordinary day's work. Extraordinary efforts, emotional excitement, sensual excesses, or malnutrition (either by insufficient food or dyspeptic habits),

induce a general lassitude—a warning that the organism is being overtasked. Repose and a healthier or more liberal diet will soon restore the functional vigor of the system. But during such periods of their diminished activity the vital powers can be rallied by drastic drugs or tonic beverages—in other words, by poisons. The prostrate vitality rises against a deadly foe, as a weary sleeper would start at the touch of a serpent; and, as danger will momentarily overcome the feeling of fatigue, the organism labors with restless energy till the poison is expelled. This feverish reaction dram-drinkers (patent dram-drinkers especially) mistake for a sign of returning vigor, persistently ignoring the circumstance that the excitement is every time followed by a prostration worse than that preceding it. Feeling the approach of a relapse the stimulator then resorts to his old remedy, thus inducing another sham revival, followed by an increased prostration, and so on; but before long the dose of the stimulant, too, has to be increased, the stimulator becomes a slave to his poison, and passes his life in a round of morbid excitements and morbid exhaustions—the former at last nothing but a feeble flickering-up of the vital flame, the latter soon aggravated by sick-headaches, "vapors," and hypochondria.

The stimulant habit in all its forms—"exhilarating beverages," "tonic medicines," "prophylactic bitters," etc.—is a dire delusion. A healthy man needs no artificial excitants; the vital principle in its normal vigor is an all-sufficient stimulus; the inspiration bought at the rum-shop is but a poor substitute for the spontaneous exaltations of a healthy mind in a healthy body. Playing with poisons is a losing game; the sweetness of the excitement is not worth the bitter reaction. In sickness stimulants can not further the actual recovery by a single hour. There is a strong progressive tendency in our physical constitution; Nature needs no prompter: as soon as the remedial process is finished, the normal functions of the organism will resume their work as spontaneously as the current of a stream resumes its course after the removal of an obstruction. A "prophylactic" brandy is old Old Scratch in the *rôle* of an exorcist. Fevers can be prevented by other means; and at any rate the possible danger of a climatic disease is preferable to the sure evils of the poison-drug. But how can noxious stimulants be distinguished from wholesome drinks? Tonic medicines, stimulating beverages, and poisons, are synonymous terms. Every known poison can become a lusted-after stimulant by forcing it repeatedly upon the (at first) reluctant stomach. It is true that the hankering of an old *habitué* after his tipple resembles the craving of a hungry man for food, but that constitutes no reproach against Nature, for the taste of the first drink betrayed the poison. To the palate of a child narcotic stimulants are bitter, alcohol is burning-acrid, tobacco nauseous, mineral poisons either bitter or insipid. By a liberal admixture of sugar and milk the repulsiveness of various narcotic decoctions can be diminished, but in no disguise could they be possibly mistaken for nourishing substances if the natural-depravity dogma had not weakened our confidence in the testimony of our instincts.

4. The Cold-Air Fallacy.—The influence of antinaturalism is most strikingly illustrated in our superstitious dread of fresh air. The air of the out-door world, of the woods and hills, is, *par excellence*, a product of Nature—of wild, free, and untamable Nature—and therefore the presumptive source of innumerable evils.

Cold air is the general scape-goat of all sinners against Nature. When the knee-joints of the young debauchee begin to weaken, he suspects he has "taken cold." If an old glutton has a cramp in the stomach, he ascribes it to an incautious exposure on coming home from a late supper. Toothache is supposed to result from "draughts"; croup, neuralgia, mumps, etc., from the "raw March wind." When children have been forced to sleep in unventilated bedrooms till their lungs putrefy with their own exhalations, the *materfamilias* reproaches herself with the most sensible thing she has been doing for the last hundred nights—" opening the windows last August when the air was so stiflingly hot." The old dyspeptic, with his cupboards full of patent nostrums, can honestly acquit himself of having yielded to any natural impulse; after sweltering all summer behind hermetically closed windows, wearing flannel in the dog-days, abstaining from cold water when his stomach craved it, swallowing drugs till his appetite has given way to chronic nausea, his conscience bears witness that he has done what he could to suppress the original depravity of Nature; only once the enemy got a chance at him: in rummaging his garret for a warming-pan he stood half a minute before a broken window—to that half-minute, accordingly, he attributes his rheumatism. For catarrh there is a stereotyped explanation : " Catched cold." That settles it. The invalid is quite sure that her cough came on an hour after returning from that sleigh-ride. She felt a pain in the chest the moment her brother opened that window. There is no doubt of it—it's all the night-air's fault.

The truth is, that cold air often reveals the existence of a disease. It initiates the reconstructive process, and thus apparently the disease itself, but there is a wide difference between a proximate and an original cause. A man can be too *tired* to sleep and too *weak* to be sick. Bleeding, for the time being, may "break up" an inflammatory disease; the system must regain some little strength before it can resume the work of reconstruction. The vital energy of a person breathing the stagnant air of an unventilated stove-room is often inadequate to the task of undertaking a restorative process—though the respiratory organs, clogged with phlegm and all kinds of impurities, may be sadly in need of relief. But, during a sleigh-ride, or a few hours' sleep before a window left open by accident, the bracing influence of the fresh air revives the drooping vitality, and Nature avails herself of the chance to begin repairs, the lungs reveal their diseased condition, i. e., they proceed to rid themselves of the accumulated impurities. Persistent in-door life would have aggravated the evil by postponing the crisis, or by turning a temporary affection into a chronic disease. But in a plurality of cases Nature will seize even upon a transient improvement of the external circumstances: a cold night that disinfects the atmosphere of the bedroom in spite of closed windows, a draught of cool air from an adjoining room, or one of those accidental exposures to wind and weather which the veriest slave of the cold-air superstition can not always avoid. For, rightly understood, the external symptoms of a disease constitute a restorative process that can not be brought to a satisfactory issue till the cause of the evil is removed. So that, in fact, the air-hater confounds the cause of his recovery with the cause of his disease. Among nations who pass their lives out-doors, catarrh and scrofula are almost unknown; not fresh air, but the want of it, is the cause of countless diseases, of fatal diseases where people are in the habit of *nailing down* their windows every winter to keep their children from opening them. "In one such den," says Dr. Bock, "I was so overcome with nausea that I could not speak till I had knocked out a pane of glass. That is about the best thing one could do in most sick-rooms"—except knocking out the whole window. The only objection to a "draught" through a defective window is, that the draught is generally not strong enough. An influx of fresh air into a fusty sick-room is a ray of light into darkness, a messenger of Vishnu visiting an abode of the damned. Cold is a disinfectant, and under the pressure of a high wind a modicum of oxygen will penetrate a house in spite of closed windows. This circumstance alone has preserved the lives of thousands whom no cough-sirup or cod-liver oil could have saved.

5. THE FEVER FALLACY.—Fever-and-ague, being eminently a summer disease, could not very well be ascribed to cold air; but the anti-naturalists, still resolved to find an extraneous cause, have selected as their scape-goat the only kind of natural food and drink most Christians ever touch in summer-time—fruits and cold water. The police of fever-stricken towns prohibit the sale of fresh fruit; fever-patients are kept in sweat-boxes, asking in vain for water and fresh air; illustrated almanacs implore us to fortify our constitutions with patent brandy—"a reliable febrifuge, and in malarious districts the only safe beverage."

Considering the problem from a purely inductive stand-point, we shall find that fruits and fevers are not necessarily concomitant. Some two hundred millions of our fellow-men stick to a frugal diet in the swampiest districts of the intertropical regions, and yet enjoy a greater immunity from periodical fevers than the inhabitants of our Northern sea-port towns. Siam, the Punjaub, the Brazilian forest-province of Entre Rios, and the swampy peninsula of Yucatan, would be the healthiest regions of this planet if the absence of what we call malarial diseases could be accepted as a safe criterion; but the accounts of former travelers show that the same diseases were entirely unknown in regions which are now justly dreaded—by visitors from the North. In the valley of the Amazon, and on the larger islands of the West Indian archipelago, fevers made their first appearance with the advent of European colonists. The natives of Sierra Leone, Dr. Schweinfurth tells us, call swamp-fever the "English sickness" —a disease confined to foreigners. The Portuguese and Italians, people with a natural predilection for a frugal diet, survive where beef-eaters die by hundreds. In Mexico, where several coast-towns have become international sea-ports, vegetarians are almost the only permanent foreign residents; native domestics, who share the flesh-pots of their foreign employers, die by scores every summer. But the necessity of such a result might have been inferred from an *a priori* axiom which seems to have been no secret to the ancient inhabitants of Southern Europe, viz., that in a warm climate calorific food is incompatible with the constitution of the human body. The word *fever* (Latin *febris*) and its equivalents in several other languages (Greek πύρεξις, Spanish and Italian *calentura*) are derived from adjectives meaning *fervid*—hot or heated—thus indicating the chief characteristic, and, according to the ancient Greek and modern Spanish theory, also the chief cause, of all pyrexial disorders. Man is a native of the tropics, and like our next relatives, the anthropoid four-handers, our primogenitor subsisted probably on fruits and water

POPULAR FALLACIES. 253

—i. e., on a refrigerating diet. In subsequent ages several tribes of the human race emigrated to regions whose climate requires calorific food and warm clothing. On returning to the birth-land of their race these wanderers often persist in habits compatible only with a low temperature : the combined influence of a warm climate, warm clothing, and calorific food overcomes the vital power of resistance; the inability of the system to preserve its due mean temperature induces the blood-changes which characterize the symptoms of climatic fevers—the overheated blood ferments. Humid heat accelerates the disintegrating process; but, that humidity is only an adjuvant and not even a necessary adjuvant cause, is proved by the immunity of fruit-eaters in the swampiest regions of the equatorial coastlands, as well as by the frequency of yellow-fever epidemics in such places as Vera Cruz and Pernambuco, whose neighborhood rivals that of Persepolis in sandy aridity. In other words, fevers are caused by the folly of aggravating the influence of the summer heat by superfluous clothing and calorific food (meat, greasy made-dishes, and ardent spirits), and not by fruit or cold water.

6. THE SPA FALLACY.—According to the theory of the anti-naturalists, a man's instincts conspire for his ruin; whatever is pleasant to our senses must be injurious; repulsiveness and healthfulness are synonymous terms. To every poison known to chemistry or botany they attribute remedial virtues; to sweetmeats, fruits, fresh air, and cold spring-water all possible morbific qualities. But, for consistency's sake, they make an exception in favor of mineral springs. Spas, impregnated with a sufficient quantity of iron or sulphur to be shockingly nauseous, **must** therefore

be highly salubrious. Solitary mountain-regions afflicted with such spas become the favorite resort of invalids; dyspeptics travel thousands of miles to reach a spring that tastes like a mixture of rotten eggs and turpentine. Faith does wonders, but the cure of a large proportion of the many thousands who annually visit such watering-places as Ems, Carlsbad, and White Sulphur Springs, need not be ascribed to the effects of imagination alone. The motion and the excitement of traveling exert a beneficial influence on many disorders. Mountain-air is almost a panacea. Woodland rambles, changes of diet and of general habits, conversation, and even music, are not unimportant co-agents of materia medica. But the spa itself—in the case of *bona fide* health-seekers, at least—is a decided drawback upon such advantages. Saline and sulphur springs are purgative; the system hastens to rid itself of an injurious substance. A very small dose might operate as a moderate aperient; but the trouble is, that the digestive organs come to rely on such excitants as they would upon alcoholic tonics, hence the chronic constipations that so often follow upon the return from a watering-place trip: the stimulant being withdrawn, the organs become remiss in their functions. From a hygienic standpoint a sanitarium without a spa is therefore by no means a Hamlet-drama minus the Prince; the mountain-air of Meran in the Tyrol or the sweet grapes of a Rhenish *Trauben-Kur* are worth a million sulphur-springs; and, if people knew half the value of up-hill pedestrian exercise, there would be a "Hygienic Home" wherever a steep mountain overlooks a populous plain.

7. THE ASCETIC FALLACY.—The origin of asceticism is widely different from that of the frugal philosophy

which consoles itself with the reflection that the reduction of our wants is equivalent to the enlargement of our means. A man of simple habits may be both happier and healthier than the lover of artificial luxuries, but the anti-naturalists make war upon earthly enjoyments as such; they try to suppress harmless as well as vicious pleasures; their aim is not the reduction but the destruction of our natural desires. The joy-loving Greeks deified even the aberrations of our natural instincts; the ascetic condemns even their legitimate gratifications. In the world of the mind as well as in the wonders of the visible creation, in streams and passions, in woods and dreams, wherever the children of Nature sought a god, the anti-naturalists feared a devil; to the exponents of asceticism life is a penalty, and earth the devil's vanity-fair, "a fleeting show, for man's illusion given." They make joy a crime, they tell us that God delights in the mortification of his creatures, in the suppression of their natural affections: "If any man hate not his father and mother and wife and children and brothers and sisters, yea, and his own life, he can not be my disciple."

But this war against Nature is the pendulum's struggle against the law of gravitation; it is the school-boy's attempt to obstruct the sources of the Danube. Swinging left, swinging right, the pendulum must return to the middle; the stream will find its way to the valley athwart all dams, in spite of all obstructions. We can not suppress the sources of a natural instinct; all we can achieve by such attempts is to divert the stream from its normal course—to turn a natural into an unnatural passion. Education, i. e., *guidance*, does not deserve its name where it is nothing but a blind struggle against Nature. Few parents know how much easier it is to *guide* than to suppress the natural propensities of a child. Obstinate vices are often merely instincts astray, perverted energies that might be made innocuous by guiding them back to their proper sphere—perverted faculties whose abuse might have been prevented by encouraging their right use. The enemies of Nature seem to believe that an instinct can be deadened by stifling its symptoms, but the history of the last eighteen centuries has demonstrated the fallacy of that principle. They tried to stop the stream: they have only succeeded in turning it from its natural course. The attempt to suppress the pursuit of natural sciences led to the pursuit of *pseudo*-sciences—to supernaturalism, demonism, and all sorts of hideous chimeras. The monastic exiles from human society peopled their solitude with phantoms. The suppression of healthful pastimes begot a passion for vicious pastimes, and made the fancied identity of sin and pleasure a sad reality. The suppression of rational freedom has led to anarchy: the pendulum swings in the opposite direction to re-establish the due equilibrium. The ordinance of celibacy became the mother of secret vices; intolerance is the parent of hypocrisy. Wherever asceticism has trampled the flowers of this earth, the soil has produced a rich crop of weeds. The pent-up well-springs of Nature have found new outlets through dark, underground currents that could not fertilize the fields, and have undermined the foundations of many useful buildings before they could regain the light of day. Whatever liberties we now enjoy had thus to force their way through unnatural obstructions, and the rise of our new civilization is merely the reappearance of a river which once flowed with a less turbulent and less turbid current. Yet it must flow on; all opposition has proved in vain, for each re-enforcement of the mole has also re-enforced the pressure of the waters.

Shall we persist in a hopeless endeavor? The dam-builders are still at work, but the rising stream surges with ominous eddies, constantly threatening to burst through all obstructions and cover the valley with wreck and ruin. There is only one remedy: We must reopen the natural channel. We must repair and improve its ancient banks—remove the dam that obstructs the stream, and build a dike along the shore.

The religion of the ancients exalted vice as well as Nature. Our present religion suppresses Nature as well as vice. The religion of the future will teach us to distinguish between vice and Nature.

THE END.

www.ingramcontent.com/pod-product-compliance
Lightning Source LLC
Chambersburg PA
CBHW021359230426

43666CB00006B/582